Green Energy and Technology

Climate change, environmental impact and the limited natural resources urge scientific research and novel technical solutions. The monograph series Green Energy and Technology serves as a publishing platform for scientific and technological approaches to “green”—i.e. environmentally friendly and sustainable—technologies. While a focus lies on energy and power supply, it also covers "green" solutions in industrial engineering and engineering design. Green Energy and Technology addresses researchers, advanced students, technical consultants as well as decision makers in industries and politics. Hence, the level of presentation spans from instructional to highly technical.

Indexed in Scopus

Indexed in Ei Compendex

Zhijun Peng
Editor

Trends and Technological Challenges in Green Energy

Selected Papers from ICGET 2023

Springer

Editor
Zhijun Peng
School of Engineering
University of Lincoln
Lincoln, Lincolnshire, UK

ISSN 1865-3529 ISSN 1865-3537 (electronic)
Green Energy and Technology
ISBN 978-3-031-55578-7 ISBN 978-3-031-55579-4 (eBook)
https://doi.org/10.1007/978-3-031-55579-4

This Springer imprint is published by the registered company Springer Nature Switzerland AG
The registered company address is: Gewerbestrasse 11, 6330 Cham, Switzerland

Paper in this product is recyclable.

Preface

The 2023 8th International Conference on Green Energy Technologies (ICGET2023) was successfully held during 17–19 August 2023 as physical conference finally after COVID-19.

The event was highlighted by keynote speakers including Prof. Dimitrios Karamanis from University of Patras, Greece, Prof. Kei Eguchi from Fukuoka Institute of Technology, Japan, and Prof. Hartmut Hinz from University of Applied Sciences, Germany. Researchers and students from Greece, Japan, Germany, Mexico, China, Ireland, UAE, Turkey, UK, Croatia, etc. have joined the conference and share their works through onsite oral sessions and online oral session.

ICGET has been an excellent forum for scientists and engineers throughout the world to present and discuss the latest technological advancements as well as future directions and trends in the field. All papers submitted to the Conference were peer-reviewed. Accepted papers are included in this volume, which comprises three chapters: (1) Solar Energy Technology and Photovoltaic Power Generation, (2) Wind Power Generation and Ocean Energy, and (3) Renewable Energy Systems and Thermal Engineering.

As this volume is published, we would like to express our sincere appreciation to all the individuals who have contributed to ICGET 2023 in various ways. Special thanks are extended to our colleagues in program committee for their thorough review of all the submissions, which is vital to the success of the conference, and also to the members in the organizing committee who had given their valuable time and efforts in planning, promoting, organizing, and helping the conference.

We warmly welcome you to join our conference next year!

Proceedings Editor

Lincoln, Lincolnshire, UK — Zhijun Peng

Committees

Conference General Chair
Hartmut Hinz, University of Applied Sciences, Frankfurt, Germany

Conference Chair
Zhijun Peng, University of Lincoln, UK

Conference Program Chairs
Bin Chen, Beijing Normal University, China
Quan Li, University of Edinburgh, UK
Xi LU, Tsinghua University, China
Dimitrios Karamanis, University of Patras, Greece

Publication Chairs
Kei Eguchi, Fukuoka Institute of Technology, Japan
Zhe Chen, Aalborg University, Denmark

Local Organizing Chairs
Schäfer Sebastian, University of Applied Sciences, Frankfurt, Germany
Carolina Tranchita, University of Applied Sciences, Frankfurt, Germany

Publicity Chair
Lihnida Stojanovska-Georgievska, University Ss Cyril and Methodius in Skopje, North Macedonia

Technical Committees
Sanjay Dey, Ananda Mohan College, India
Kreangkrai Maneeintr, Chulalongkorn University, Thailand
Chitchai Srithapon, KTH Royal Institute of Technology, Sweden
Joseph K. Bwapwa, Mangosuthu University of Technology, South Africa
Ghulam Murshid, Sultan Qaboos University, Muscat, Oman
Feng Hu, Chinese Academy of Sciences, China
Syazwani Idrus, Universiti Putra Malaysia, Malaysia
Mohd Azlan Musa, University Malaysia Terengganu, Malaysia

Contents

Part I Solar Energy Technology and Photovoltaic Power Generation

Performance Analysis for Bidirectional DC/DC Power Converter for Photovoltaic Applications: Simulation Study 3
Sasa Sladic, Michele De Santis, Even Zivic, and Wojciech Giernacki

Urban Transformation to Carbon-Free with Lush Greenery and Colored Solar Energy and Storage Technologies at the Diverse Climatic Conditions of Europe 11
D. Karamanis, H. -Y. Liu, D. Avisar, L. Braslina, L. F. Cabeza, D. D'Agostino, V. Kapsalis, P. Lapka, O. Raita, N. Skandalos, and F. Vanhuyse

South Africa's Solar Rebates: Hope or Hate for Residential Customers Within eThekwini Municipality 25
Leshan Moodliar and Innocent Ewean Davidson

A Review of LED Driver Topologies and Control Methods for Energy-Efficient Smart Farming Application 37
M. L. Pay, J. Christensen, F. He, L. Roden, M. Rasel, L. Tyler, H. Ahmed, and M. Foo

Part II Wind Power Generation and Ocean Energy

Monitoring Wind Turbine Blade Using Interferometric Radar 49
Gokhan Kilic

Generation of Fluid Field Around Tidal/Ocean Power Generator with Unidirectional Flow Exploiting Variable Auto Encoder 59
Ryo Fujiwara, Hiroki Nakagawa, Tsubasa Ebiko, and Makoto Miyatake

Finite Element Analysis of Large-Span Valve Hall Structure Under Extreme Wind Conditions 73
Dongmei Fan, Wendou Yan, Yi Liang, Weiyuan He, and Gang Lin

Part III Renewable Energy Systems and Thermal Engineering

Dynamics of Renewable Energy Penetration for the Malaysian Electricity Generation 85
Zong Qing Yeoh, Andrew Huey Ping Tan, Anwar P. P. Abdul Majeed, Wei Chen, and Eng Hwa Yap

Optimal Sizing and Operation of Hybrid Renewable Energy System: A Techno-Economic Analysis of Residential Building Considering Seasonal Storage 99
Kaichen Qu and Hong Zhang

Numerical Simulation About Transient Performance of Regenerative Cooling System for Combustion Chamber 113
Chen Zhang, Hui Gao, Jiajun Zhao, and Dongsheng Wen

An Evolutionary Game Study on the Behavioral Strategies of Carbon Label Stakeholders Based on Prospect Theory 125
Chaoping Liu and Xiaomei Wu

Index 141

Part I
Solar Energy Technology and Photovoltaic Power Generation

Performance Analysis for Bidirectional DC/DC Power Converter for Photovoltaic Applications: Simulation Study

Sasa Sladic, Michele De Santis, Even Zivic, and Wojciech Giernacki

1 Introduction

Bidirectional power converters can be found in literature very often [1–12]. Bidirectional DC/DC power converters can be combined with bidirectional DC/AC converters, boosting the importance of DC/DC power converters both to AC and DC electrical networks (microgrids). Due to the large number of different DC/DC bidirectional converters, most of the authors limit their analysis to transformerless bidirectional DC/DC converters [1–6].

Others try to improve the performance of known bidirectional DC/DC power converters [7], improve their efficiency [8], or develop new topologies in order to compare them to known ones [9].

The most frequent topologies include buck–boost topology and symmetrical buck–boost topology [1–5]. Buck–boost topology consists of two power converters (single buck and single boost). Symmetric power converters consist of two power converters of the same type (e.g., buck–boost), which correspond to H-bridge topology [6]. More unusual combinations (e.g., zeta/SEPIC) mean that it is theoretically possible to combine all the power DC/DC converters [9]. Because of the importance of PV systems in the last 20 years, a huge effort was taken for the

S. Sladic (✉)
Faculty of Engineering, University of Rijeka, Rijeka, Croatia
e-mail: sladics@riteh.hr

M. De Santis
Cusano University, Rome, Italy

E. Zivic
Faculty of Engineering, University of Pula, Pula, Croatia

W. Giernacki
Institute of Control, Robotics and Information Engineering, University of Poznan, Poznan, Poland

Z. Peng (ed.), *Trends and Technological Challenges in Green Energy*, Green Energy and Technology, https://doi.org/10.1007/978-3-031-55579-4_1

classification of DC/DC power converters for renewable energy source applications [10–12]. In this chapter, Luo [13] and flyback power converters have been combined. Their combination is expected to improve DC/DC bidirectional converter efficiency.

Most of the DC/DC power converters have been known for decades. However, Luo power converter has only been presented recently [13]. Basically, the main idea for the new class of step-up power converters was developed in the last 20 years. Luo converter is known for its computational complexities [14]. It could be said that it incorporates properties of basic (e.g., boost power converter) with the addition of switched capacitor converters (SCCs). This means that Luo converters are rich in complex behavior and difficult to analyze. However, it has numerous advantages, including a high step-up DC–DC conversion ratio and an increased efficiency compared to the boost power converter. In spite of the obvious advantages, its modified topologies are hard to recognize between the most used bidirectional power converters.

Besides, Luo converter has a larger ratio between input and output voltage, which is useful in PV systems [15]. During the last 15 years, battery voltage in electric vehicles has increased, so a large step-up ratio ha become important in vehicle chargers [16]. Similar topologies can be found in the literature [17], which means Luo converter is still being explored widely.

2 Luo-Based Topology

The Buck–boost bidirectional power converter shown in Fig. 1 has been used most frequently. This type of power converter can connect two energy storages (e.g., supercapacitor and batteries).

Its alternative is the Luo-based bidirectional DC/DC power converter (Fig. 2). This type of power converter has an additional capacitor C, two diodes (D_3 and D_4), and an additional transformer. However, this bidirectional power converter has both transistors on the low potential. This means that low potential (boost) driver can be used for switching of both transistors. Furthermore, Luo converter step-up ratio brings to this converter a higher ratio between input and output voltage (both inputs could be input and output) and increased efficiency. Numerous simulations were taken.

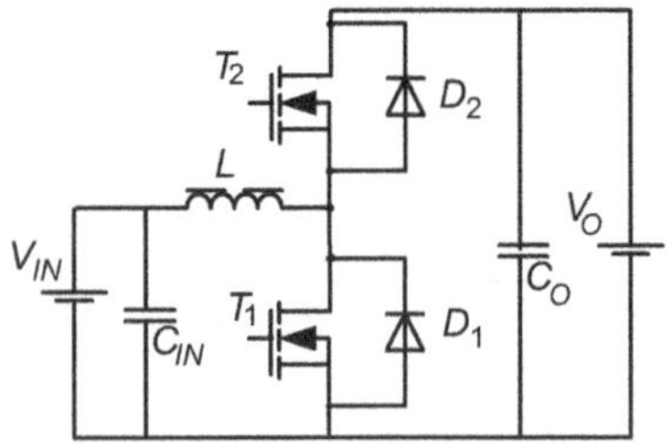

Fig. 1 Buck–boost (standard) bidirectional DC/DC power converter

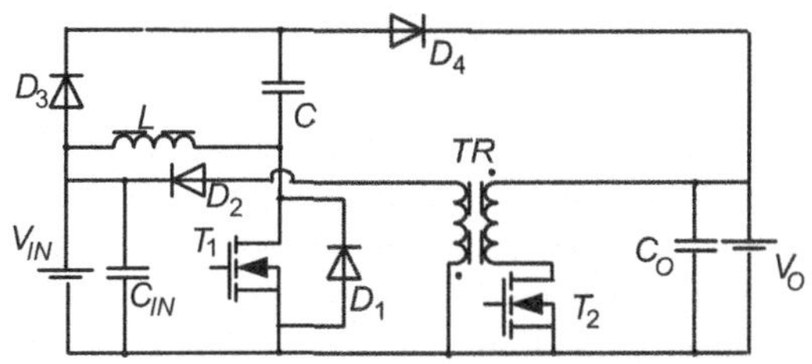

Fig. 2 Luo-based (proposed) bidirectional DC/DC power converter

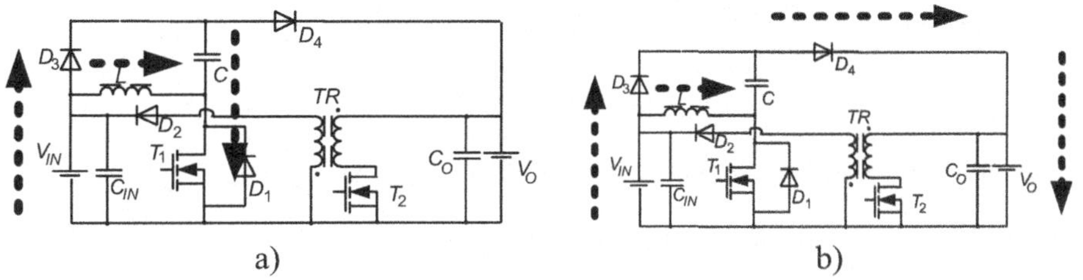

Fig. 3 Luo converter operation in the frame of bidirectional Luo-based converter: (**a**) T_1 ON; (**b**) T_1 OFF

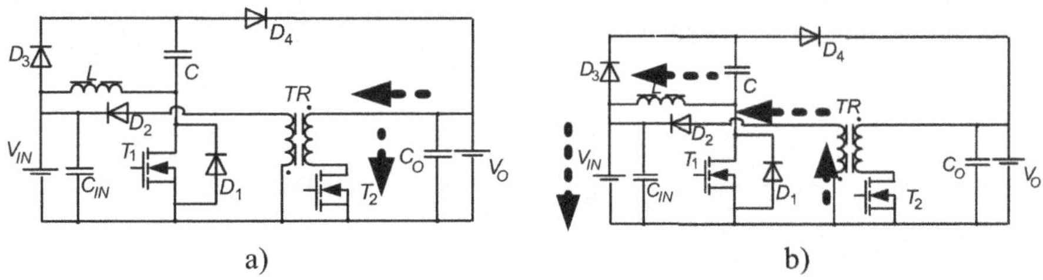

Fig. 4 Flyback converter operation in the frame of bidirectional Luo-based converter: (**a**) T_2 ON; (**b**) T_2 OFF

In this power converter, input voltage V_{IN} was increased via action of boost converter (V_{IN}, L, T_1, D_2). This means that the energy stored in L has to be released after the switching of the transistor T1 (Fig. 3). The input voltage supported with induced voltage on inductance L is charging output capacitor CO. In this case, energy from the input battery and input capacitor (C_{IN}) is transferred to the output.

In order to obtain the opposite energy direction (C_O to C_{IN}), flyback converter (T_2) action should occur (Fig. 4). After switching ON the transistor T_1, magnetic field increases in transformer TR. This means that after turning OFF the switch T_1, magnetic field energy is released and diode D_2 conducts current, which is charging capacitor C_{IN}.

2.1 Luo Versus Boost Topology

In order to analyze the advantages of the proposed topology, it is necessary to compare boost and Luo converters.

Luo converter, in its basic form, has an additional capacitor and a diode. However, as the main energy storage, capacitor C_1 has been used. A combination of inductances and capacitances results in more complex behavior of the converter (Fig. 5). Assuming that capacitor C_1 is large enough, an equation for the output voltage V_O can be written as

$$V_O/V_{IN} = (2 - D)/(1 - D) \tag{1}$$

Compared to boost, Luo converter has a higher output voltage:

$$V_{IN}/V_S = 1/(1 - D) \tag{2}$$

In other words, the same output voltage could be obtained with lower duty cycle D and lower inductance current. Since inductance current is connected with parasitic resistance in inductance winding, Luo converter has a higher efficiency.

According to this equation, it can be noted that at the same duty cycle and same input voltage, the output voltage for the Luo converter will be higher than the boost

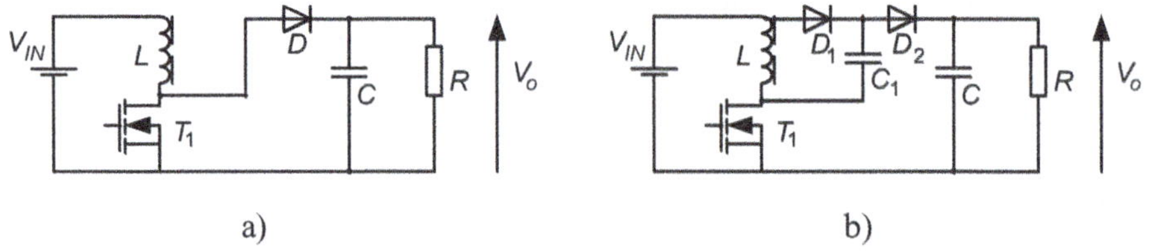

Fig. 5 (**a**) Boost converter; (**b**) Luo converter

Fig. 6 Output voltage (V_O) simulated for boost and Luo converter under the same conditions (1 V/div., 20 ms/div.)

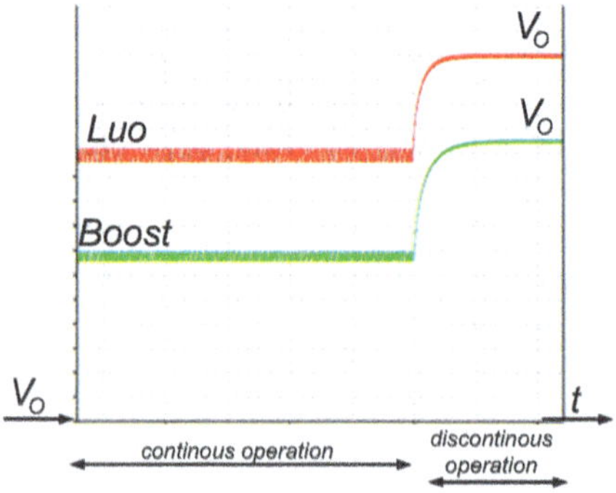

power converter. This conclusion is valid both for continuous and discontinuous operation modes (Fig. 6).

2.2 *Reversing the Direction of the Energy Flow in Bidirectional Luo-Based Converter*

The direction of energy flow depends on which transistor is switching. In case of Luo converter action, T_1 is switching, and in case of flyback converter action, T_2 is switching. This means that precise control of flow direction is possible and even break of current flow (diodes D_2 and D_4 are not conducting). Figure 7 shows the action of both converters. Figure 8 shows the detail during the energy direction change.

Fig. 7 Luo converter current (i_1) and flyback converter current (i_2) with associated PWM signals (1 A/div., 7.5 V/div., 200 ms/div.)

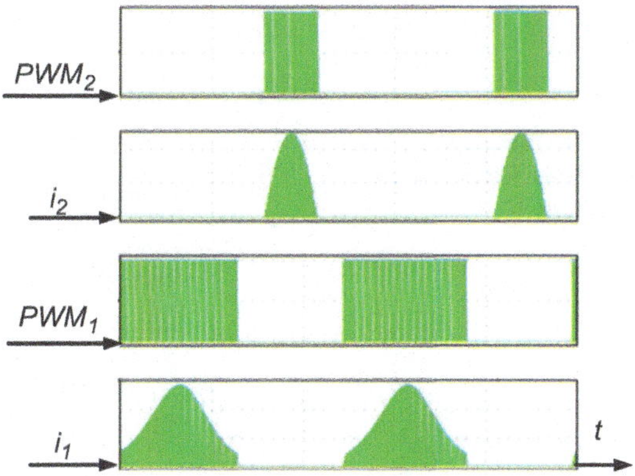

Fig. 8 Luo converter current and flyback converter current with associated PWM signals (1 A/div., 7.5 V/div., 20 ms/div.)

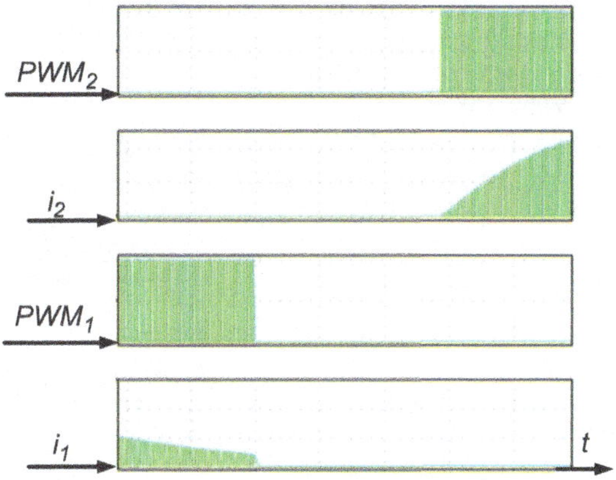

3 Bidirectional Power Converter in Photovoltaic Applications

In photovoltaic power systems, bidirectional power converter has been used in combination with no consistent power source. Combinations of batteries and supercapacitors, solar cells and batteries, supercapacitors and fuel cells, and other combinations can be used.

Bidirectional power converter in photovoltaic systems can be controlled in such a way that during high intensity of solar radiation (increased solar array voltage) battery is charged via bidirectional power converter action. Otherwise, when the solar radiation is weak (decreased solar array voltage) battery is discharged in order to give enough power to (step-up) power converter (Fig. 9).

In contrast to the earlier case (Fig. 8) where the action of one converter was replaced with the action of other converters periodically, in this case (Fig. 9), it is a consequence of the solar radiation. Also, more irregular transitions can occur.

Figure 10 shows the case where the majority of the time Luo converter has been active. This means that solar irradiation is high enough and battery is charging. However, during the short portion of time (reduced solar irradiation), the energy generated from the solar array is not enough. The control circuit (Fig. 9) detected

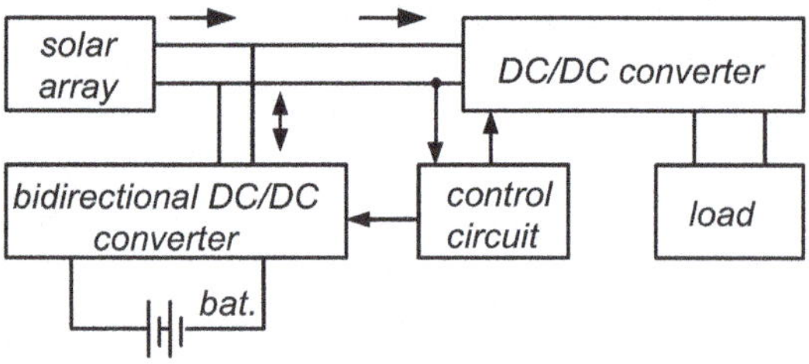

Fig. 9 Bidirectional power converter for photovoltaic applications

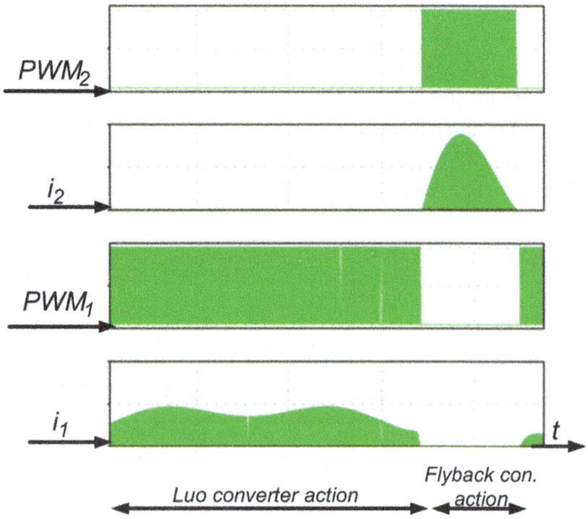

Fig. 10 Luo converter action and short interval flyback converter action in the frame of bidirectional power converter (0.5 A/div., 7.5 V/div., 200 ms/div.)

reduced power (reduced current or voltage) and instead of the signal on transistor T_1 (PWM_1) PWM signal on another transistor appears (PWM_2). At this moment, the battery starts to discharge. Charging and discharging of battery can be defined precisely without overlapping.

4 Conclusions

A simulation study of Luo-based bidirectional power converter was done. It has been shown that this type of power converter can deliver better performances compared with standard boost–buck power converter. Its advantages are that booth power transistors are on low potential, enabling the usage of cost-effective MOSFET/IGBT drivers. A higher ratio between output and input voltage gives the Luo converter some advantage. The same output ratio can also be accomplished with a lower duty cycle, which means lower dissipation on the Luo converter inductance. In other words, Luo converter has already an advantage over boost power converter regarding the efficiency; however, buck converter has a robust construction and is known for high efficiency, so the overall study of bidirectional power converter comparison demands operating instead of simulation models. Following the simulation results analyzed, Luo-based topology would be more appropriate for cases where the difference between the input and output voltage is high and boost converter could hardly be used (e.g., hybrid vehicles and photovoltaic systems).

References

1. Alexander Wallberg, Design and Construction of a Bidirectional DC/DC Con-verter, Uppsala University, June 2019. https://www.diva-portal.org/smash/get/diva2:1323585/FULLTEXT01.pdf
2. Nils Johansson, Erik Jarnfors, Bidirectional DC/DC converter for application in an electric vehicle for better battery management, Division of Industrial Electrical Engineering and Automation Faculty of Engineering, Lund University
3. Junhong Zhang, Bidirectional DC-DC Power Converter Design Optimization, Modeling and Control,
4. TI Designs: TIDA-01168 Bidirectional DC-DC Converter Reference Design for 12-V/48-V Automotive Systems, Texas Instruments, March 2018.
5. M. Ejas Ahamed, S. Senthil Kumar, Review of Bidirectional DC-DC Converters, International Journal of Advance Research and Innovation, Vol 5(4), Oct–Dec 2017
6. Boris Axelrod, Yefim Berkovich, Yuval Beck, Family of universal bidirectional DC–DC converters with an extended voltage gain, IET Power Electronics, September 2019 https://ietresearch.onlinelibrary.wiley.com/doi/epdf/10.1049/iet-pel.2018.6243 modified H bridge
7. Hsuan Liao, Yi-Tsung Chen, Linda Chen, Jiann-Fuh Chen, Development of a Bidirectional DC–DC Converter with Rapid Energy Bidirectional Transition Technology, Energies MDPI, June 2022.
8. Bor-Ren Lin, Analysis and Implementation of a Bidirectional Converter with Soft Switching Operation, Processes, MDPI, 13th March 2022.

9. Kim, I.D., Paeng, S.H., Ahn, J.W., et al.: 'New bidirectional ZVS PWMSEPIC/Zeta DC–DC converter'. IEEE Int. Symp. Industrial Electronics, Vigo,Spain, 2007, pp. 555–560
10. Kuei-Hsiang Chao, Ming-Chang Tseng, Chun-Hao Huang, Yang-Guang Liu, Liang-Chiao Huang, Design and Implementation of a Bidirectional DC-DC Converter for Stand-Alone Photovoltaic Systems, International Journal of Computer, Consumer and Control (IJ3C), Vol. 2, No.3 (2013)
11. D. Dong, F. Luo, X. Zhang, D. Boroyevich, P. Mattavelli, Grid-Interface Bidirectional Converter for Residential DC Distribution Systems-Part 2: AC and DC Interface Design With Passive Components Minimization, in IEEE Transactions on Power Electronics, vol. 27, no. 4, pp. 1667–1679, April 2013, https://doi.org/10.1109/TPEL.2012.2213614.
12. N. Kondrath, Bidirectional DC-DC converter topologies and control strategies for interfacing energy storage systems in microgrids: An overview, 2017 IEEE International conference on Smart Energy Grid Engineering (SEGE), Oshawa, ON, Canada, 2017, pp. 341–345, https://doi.org/10.1109/SGE.2017.8052822.
13. Fang Lin Luo, Luo-Converters, a series of new DC-DC step-up (boost) conversion circuits, Proceedings of Second International Conference on Power Electronics and Drive Systems, Singapore, 1997, pp. 882–888 vol.2, https://doi.org/10.1109/PEDS.1997.627511.
14. Zirui Jia, Chongxin Liu, A modified modeling and dynamical behavior analysis method for fractional-order positive Luo converter, PLoS ONE 15(8): e0237169. Doi:10.1371. Long Wang, University of Science and Technology Beijing, China, August 14, 2020
15. Kishore, P.S.V., Bogineni, J., Rajesh, J., Halder, S., Jayaram, N. (2022). Luo Converters for Renewable Energy Applications. In: Priyadarshi, N., Bhoi, A.K., Bansal, R.C., Kalam, A. (eds) DC-DC Converters for Future Renowable Energy Systems. Energy Systems in Electrical Engineering. Springer, Singapore. https://doi.org/10.1007/978-981-16-4388-0_7 Published: 28th September 2021
16. U. Sharma, B. Singh, A Multistep Bidirectional Onboard Charger for Light Electric Vehicles Using Interleaved Luo Converter, 2021 International Conference on Sustainable Energy and Future Electric Transportation (SEFET), Hyderabad, India, 2021, pp. 1–5, https://doi.org/10.1109/SeFet48154.2021.9375736.
17. Saravanan, S., Usha Rani, P. A Novel High Gain Interleaved Luo DC-DC Converter with Cascaded Fuzzy Logic Controller for PV system, Application. J. Electr. Eng. Technol. 18, 925–940 (2023). https://doi.org/10.1007/s42835-022-01279-w, March 2023

Urban Transformation to Carbon-Free with Lush Greenery and Colored Solar Energy and Storage Technologies at the Diverse Climatic Conditions of Europe

D. Karamanis, H. -Y. Liu, D. Avisar, L. Braslina, L. F. Cabeza, D. D'Agostino, V. Kapsalis, P. Lapka, O. Raita, N. Skandalos, and F. Vanhuyse

D. Karamanis (✉)
Department of Environmental Engineering, University of Patras, Agrinio, Greece
e-mail: dkaraman@upatras.gr

H. -Y. Liu (✉)
Department of Environmental Impacts and Sustainability, NILU – The Climate and Environmental Research Institute NILU, Kjeller, Norway
e-mail: hyl@nilu.no

D. Avisar
Department of Applied Mathematics, Israel Institute for Biological Research, Ness-Ziona, Israel

L. Braslina
Faculty of Business, Management and Economics, University of Latvia, Riga, Latvia

L. F. Cabeza
GREiA Research Group, Universitat de Lleida, Lleida, Spain

D. D'Agostino
European Commission, Joint Research Centre (JRC), Ispra, VA, Italy

V. Kapsalis
School of Mechanical Engineering, Sector of Industrial Management and Operational Research, National Technical University of Athens, Athens, Greece

P. Lapka
Institute of Heat Engineering, Faculty of Power and Aeronautical Engineering, Warsaw University of Technology, Warsaw, Poland

O. Raita
National Institute for Research and Development of Isotopic and Molecular Technologies, Cluj-Napoca, Romania

N. Skandalos
Czech Technical University in Prague, University Centre for Energy Efficient Buildings, Buštěhrad, Czech Republic

F. Vanhuyse
Stockholm Environment Institute, Stockholm, Sweden

Z. Peng (ed.), *Trends and Technological Challenges in Green Energy*, Green Energy and Technology, https://doi.org/10.1007/978-3-031-55579-4_2

1 Introduction

After the Paris Agreement aiming at keeping the increase of global average temperature below 1.5 °C above preindustrial levels, 194 states and the European Union, responsible for more than 98% of global greenhouse gas emissions (GHGs), have started to undertake ambitious efforts to reduce emissions, especially China (~20%) and the United States (~18%), the countries with the first and second largest CO_2 emissions among the agreement ratified or acceded members. The European Union set ambitious targets with Green Deal, aiming to achieve climate neutrality by 2050 with an economy of net-zero greenhouse gas emissions. Is this climate neutrality feasible by the middle of the century? Economically, the expenditures in systems for the energy transition account for around $275 trillion in the period 2021–2050 or $9.2 trillion per year on average [1]. The increased amount of $3.5 trillion needed from 2022 spending is roughly equivalent to half of global corporate profits in 2020 or 25% of total tax revenue or 7% of household spending. Although the required investments are substantial and many will bring logical profits, the avoided costs due to climate change physical risks are estimated to be many times more. The final costs are expected to be much less due to technological innovation that will eventually reduce the transition overall needed investment. Therefore, the energy transition can be both an inarguable and global beneficial choice to balance the earth's energy fluxes, keep average temperature increase lower than 1.5 °C, and avoid climate change-devastating consequences and an opportunity to reorganize the unprecedented growth of developed countries and support the sustainable growth of developing countries. Moreover, colonization strategies followed in the fossil fuels era will not be needed and citizens will be brought to the center of the transition.

In this global requisite boldness and resolve to broaden the climate change response urgently, cities play an important and key role since urban activities and components such as transport and buildings are the major sources of GHG emissions. Estimates from several sources and at different scales suggest that cities are responsible for 75% of global CO_2 emissions [2]. Moreover, urbanization is increasing, especially in developing countries, with a direct rising effect on the GHG emission from 25 $GtCO_2$-eq in 2015 to 29 $GtCO_2$-eq in 2020 (or an increase of global share from 62% to 67–72%) [3]. Normalizing the urban greenhouse gas emissions per capita, the trend is also increasing on average globally and reached 6.2 in 2015 from 5.5 tCO_2-eq per person in 2000 [3]. While emissions increased per capita for most of the developing countries, it only slightly decreased from 11.4 to 10.7 tCO_2-eq per person for the developed countries. According to the IPCC, aggressive and accelerated mitigation actions are needed to limit global warming to 1.5 °C with no or limited overshoot by the end of the century with electrification, energy and material efficiency, preference for renewables applications, and behavioral change. In this context, urban greenhouse gas emissions could be maximized at 3 $GtCO_2$-eq in 2050 and reach net-zero balanced by natural sinks such as greenery. Moreover, the general methodologies of carbon-neutral cities need to be adjusted to local conditions and particular needs through the interconnection of the cities' ambitious goals toward the broader scale of energy transition and climate change mitigation [4]. More clarity is needed on the definition of the conceptual approach to carbon neutrality and particularly on the role of offsetting the emissions.

Is this goal of reaching net zero sufficient at the city level? In 2022, 96% of Europe's urban population was exposed to unsafe levels of air pollution such as fine particulate matter. According to the European Environment Agency, air pollution kills more than 1200 children and teenagers per year in Europe [5]. Moreover, the assumption of biomass carbon neutrality has caused a further increase in wood burning for heating purposes in several European cities. In the European Union, fuel combustion in the residential sector is an important source of fine particulate particles (PM2.5 and PM10) and black carbon emissions contributing 56%, 40%, and 46% for the commercial, institutional, and household sectors, respectively [6]. Wood and fossil fuels burning for heating purposes in different European cities contribute significantly to the increase in ambient particulate matter levels in urban air during the wintertime and impact human health.

Given the severity of air pollution, there is an urgent need to address human exposure and health effects from elevated levels of pollutants mainly originating from the combustion of fossil fuels [7]. Recent studies indicate that a 100% zero carbon emission electricity transition is feasible and has already been attained by external renewables in several cities [8]. However, there are limited studies on cities' full decarbonization. Therefore, the carbon-neutral city concept should be advanced to carbon-free urban transformation with specific legislating implementation in the few upcoming years. Moreover, the primary energy consumption is based on buildings and transport. Since the reduction of unnecessary movements depends on society transformation into a more environmental-friendly way of living that cannot be fast due to inertia reasons, solar energy and storage applications for the built and urban environments should be accelerated toward the elimination of the use of fossil fuels and city decarbonization.

In this context, nature-based solutions (NBS) are well-accepted and developed concepts as forefront approaches to both preserve and restore ecosystem integrity and increase climate resilience. The increase of the city's park access and green space has been found beneficial for the risk reduction of all-cause, circulatory system-caused, and stroke-caused mortalities [9], supports health equity [10], and should be considered a strategic public health intervention [11]. Reconciling city economic efficiency with optimized combination of NBS and the solar energy technologies, social inclusion in a healthy city life with dignity for all and environmental responsibility with clean urban air are essential for confronting climate change toward a sustainable global prosperity.

In view of the above, the primary objective of this work is to advance the carbon-neutral city concept to carbon-free with innovative NBS and colored building integrated photovoltaics (BIPV) in a novel multipillar framework and optimize these combined central solutions for the different climatic conditions of Europe linked with urban storage through the water–energy nexus and cloud information technologies.

2 A Novel Framework for Carbon-Free Cities

According to the United Nations Framework Convention on Climate Change (UNFCCC), a carbon-neutral city is characterized as such when it has reduced the majority of carbon emissions and offset residual or remaining unavoidable emissions, to the point that its operations do not result in a net increase in emissions. C40, a global climate leadership group of 96 cities, defines carbon-neutral cities with a net-zero carbon footprint, balancing their GHGs with carbon removal or offsets. Efforts are driven to minimize emissions by using renewable energy sources, improving energy efficiency, promoting low-carbon transportation and consumption, and enhancing natural carbon sinks. The International Council for Local Environmental Initiatives (ICLEI) was founded in 1990 to support sustainable development through the design of five pathways of low emission, nature-based, equitable, resilient, and circular development to create systemic change. Climate neutrality in ICLEI is defined as the drastic reduction and sequester of GHGs and their avoidance in local, regional, and community operational activities with divest repurpose, and reinvest to an absolute net-zero emission level at the latest by 2050. GHG emissions that are not immediately removed, reduced, or avoided can be offset and compensated.

The EU Covenant of Mayors for Climate & Energy was established as an initiative to implement climate actions in local governments, voluntarily. Since its launching in 2008, more than 10,000 members have committed to achieving and exceeding the EU climate and energy targets. The covenant is based on reducing GHG emissions by 55% by 2030, strengthening resilience and alleviating energy poverty. Currently, an enormous number of best-practice actions are available (more than 1800) and action plans (7800).

Recently, the EU mission of 112 climate-neutral and smart cities was launched by the European Commission [12]. With an ambition to reach net-zero GHGs by the year 2030, the mission targets the reduction of emissions from (a) stationary energy (combustion of fossil fuels in all buildings and facilities within the city boundary, (b) vehicles and transport (combustion of fossil fuels), (c) electricity and district energy for heating/cooling from power plants within or outside the city boundary, (d) waste generated or managed within or outside the city, (e) land-use change within the city, and (f) chemical processes in the industry within the city boundary [13]. These emissions are related to the three scopes of the global protocol for community-scale greenhouse gas inventories accounting and reporting standard for cities that was developed by the partnership of World Resources Institute, C40 cities, and ICLEI networks (Fig. 1). Although an important initiative toward carbon neutrality, the implementation routes are mainly concentrated on Scope 1 without clean energy distinction of zero emissions at the interface of Scope 2 and no inclusion (yet) of consumption-based emissions inside the boundary from emissions generated outside the city boundary [14].

In summary, a plethora of different networks have been initiated at different scales toward the goal of carbon neutrality to reduce the impact of human activities

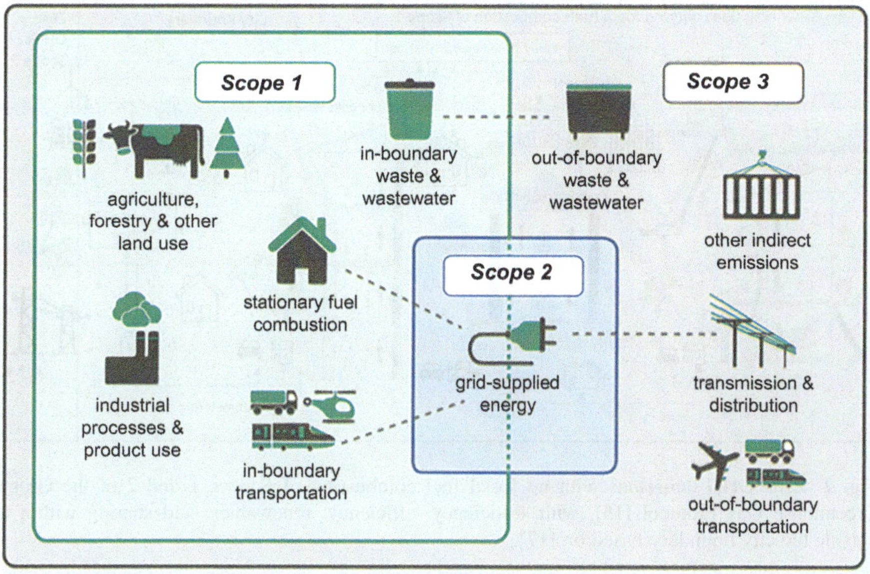

Fig. 1 Greenhouse gas (GHG) emissions within or outside the city boundary [15]

on the environment and climate change. However, there are different ambition levels in the energy transition as carbon-neutral, net-zero energy, carbon-free, circular, regenerative [16]. A carbon-neutral city in the existing networks is a city that has achieved net-zero carbon dioxide emissions, which means that the city has either eliminated its carbon dioxide emissions or offset them by other means. Reaching carbon neutrality with a net-zero energy system is already a big step for the city planning but it may be accomplished without the phase out of fossil fuels. Due to the air quality problems described above, more ambitious targets to zero GHG emissions are required at the city level beyond 2030, especially on Scope 2 of the GHG protocol with city energy use produced by 100% renewable energy either in the city or outside the boundary. With the advancing of solar city applications, the implementation of actions such as positive energy buildings, neighborhoods and districts, and new carriers such as hydrogen, full decarbonization and carbon-free cities need to be targeted with a new framework of specific indicators, beyond the current European carbon emission targets and the offset of capturing, storing, and utilizing or compensating by emission trade systems or tree planting. In this frame, we have recently shown [17] that Scope 2 can be substituted by the implementation of the SERAS (sufficiency, efficiency, renewables, and sharing) principle within or outside the city boundary (Fig. 2). In this scheme, the transmission and distribution emissions of Scope 3 are also eliminated.

In this context, the carbon-free urban transformation can be implemented by advancing the ten approaches drawn on global best-practice and case studies from

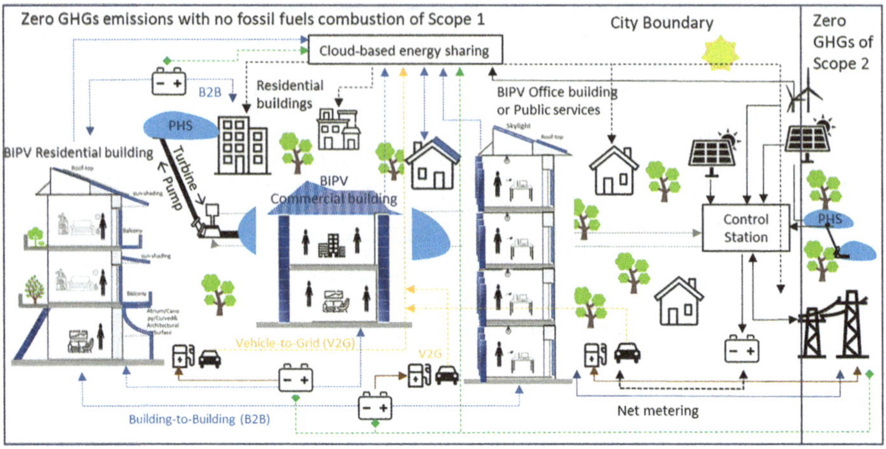

Fig. 2 Zero GHG emissions with no fossil fuel combustion of Scopes 1 and 2 of the Global Greenhouse Gas Protocol [15], with sufficiency, efficiency, renewables, and sharing within or outside the city boundary based on [17]

cities across the C40 network [18] into a new framework with actions organized into four pillars as shown in Fig. 3. The framework can be organized on multigovernance decision and implementation with top-down strategies to facilitate and support urban transformation with activities enhancing public utilities and innovative NBS and BIPV as combined central solutions (linked with IT, urban storage, and water–energy nexus), bottom-up solar citizens participation with a colored BIPV bioclimatic design, and implementation through communities and hybrid participatory co-creation with ICT tools for social innovation and awareness, integrated with PV recycling and actions for circularity, upcycling, and recycling.

3 NBS and Greenery for Beautiful, Resilient, and Healthy Cities

In recent years, there has been a growing recognition of the importance of NBS in creating beautiful, resilient, and healthy cities. Implementing NBS in cities can have numerous benefits in the pursuit of carbon-free cities [19]. These solutions contribute to (1) mitigating the urban heat island effect and cooling the environment by evapotranspiration. Studies have shown NBS effectiveness in all the climatic zones from warm Mediterranean in Barcelona to continental London with reduction of surface temperatures by up to 8 °C [20]. (2) Improving air quality since vegetation is capturing and filtering pollutants, thus reducing the levels of PM and harmful gases (up to 1.36 Mtons of air pollutants per year [21]. (3) Enhancing biodiversity with green spaces and urban forests providing habitats for a diverse range of plant

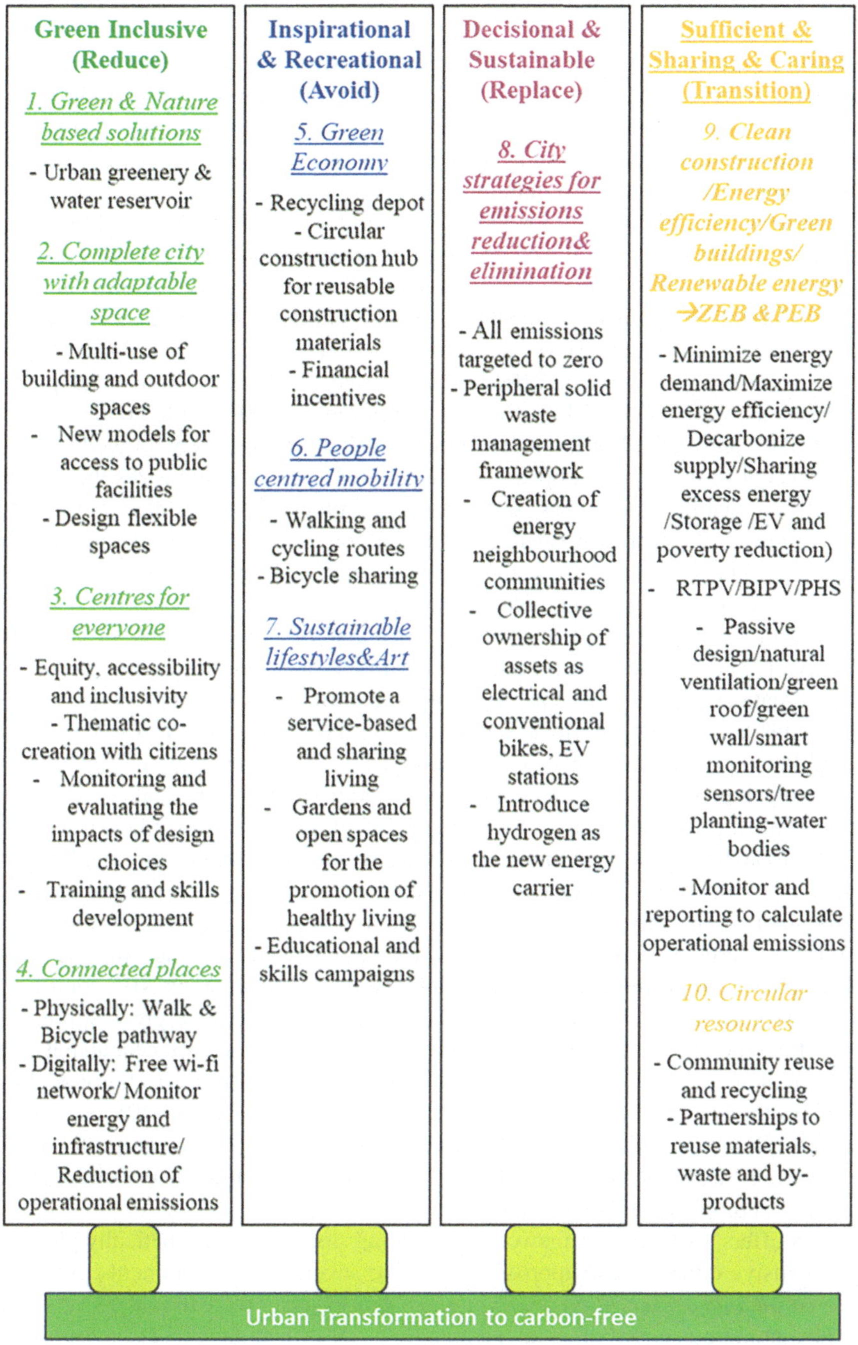

Fig. 3 A new framework for urban transformation to carbon-free by advancing the ten approaches drawn on global best-practice and case studies from cities across the C40 network [18], with greenery, rooftop photovoltaics (RTPV), and colored building integrated photovoltaics (BIPV) for zero-energy and positive energy buildings (ZEB & PEB) and storage flexible intermediates between electricity generation and supply as urban pumped hydrostorage (PHS)

and animal species, contributing to urban biodiversity. A study conducted in Berlin found that urban forests can host up to 2020 plant species, supporting a rich ecosystem [22]. (4) Positive effects on well-being and mental health as analyzed in the introduction. By incorporating nature into urban environments, cities can promote healthier and happier communities.

In Europe, the EU has acknowledged the significance of NBS in tackling environmental challenges and advancing sustainable development [19]. The European Commission (EC) has played a crucial role in supporting NBS projects through various funding programs such as the Horizon 2020, BiodiERsA ERA-Net, Biodiversity Partnership, and Horizon Europe initiatives. NBS contributes to the European Green Deal priorities, Biodiversity Strategy for 2030, Climate Adaptation Strategy, as well as the EU's Climate Ambition for 2030 and 2050 to foster biodiversity and make Europe more climate resilient. It is closely aligned with the core concept of the New European Bauhaus (NEB), which emphasizes the integration of sustainability, aesthetics, and inclusivity in the built environment. The NEB aims to bridge the world of science and technology, art, and culture, and is about leveraging our green and digital challenges to transform our lives and society through holistic solutions to global challenges.

It becomes important to analyze the potential of NBS in view of the NEB initiative [23] and conceptualize and demonstrate how to link these two approaches as schematically shown in Fig. 4, avoid tradeoffs, and enhance synergies and complementarities, through local demonstration. NBS links to the core concept of the NEB in several key aspects: (1) *sustainability*: NBS prioritizes sustainability by leveraging the benefits of nature to address environmental challenges. The NBS aligns with the United Nations Sustainable Development Goals (SDGs) and contributes to achieving targets related to climate action, sustainable cities, and responsible consumption and production. (2) *Aesthetics*: NBS integrates natural elements and aesthetics into urban design, creating visually appealing and harmonious environments. Vertical gardens, green roofs, and living walls enhance the beauty of urban spaces and promote a connection with nature. (3) *Inclusivity*: NBS contributes to inclusive urban environments by providing equitable access to green spaces and promoting social cohesion. By incorporating NBS, cities can foster inclusivity and ensure that all citizens, regardless of their socioeconomic status or background, have access to the benefits of nature. (4) *Innovation and collaboration*: NBS encourages interdisciplinary collaboration and innovation by bringing together professionals from various fields such as architecture, urban planning, ecology, and social sciences to address complex societal challenges.

NBS offers a valuable framework for creating sustainable, aesthetically pleasing, and inclusive cities. These approaches involve several key components, including (1) *creation and preservation of urban green spaces*: offer recreational opportunities but also numerous environmental advantages as described above. (2) *Incorporating green infrastructure into urban built environments*: Natural features such as green roofs, rain gardens, and permeable pavements can be used to manage stormwater, reduce flooding, and improve water quality. By mimicking natural processes, green infrastructure helps decrease the burden on traditional stormwater systems, reduce

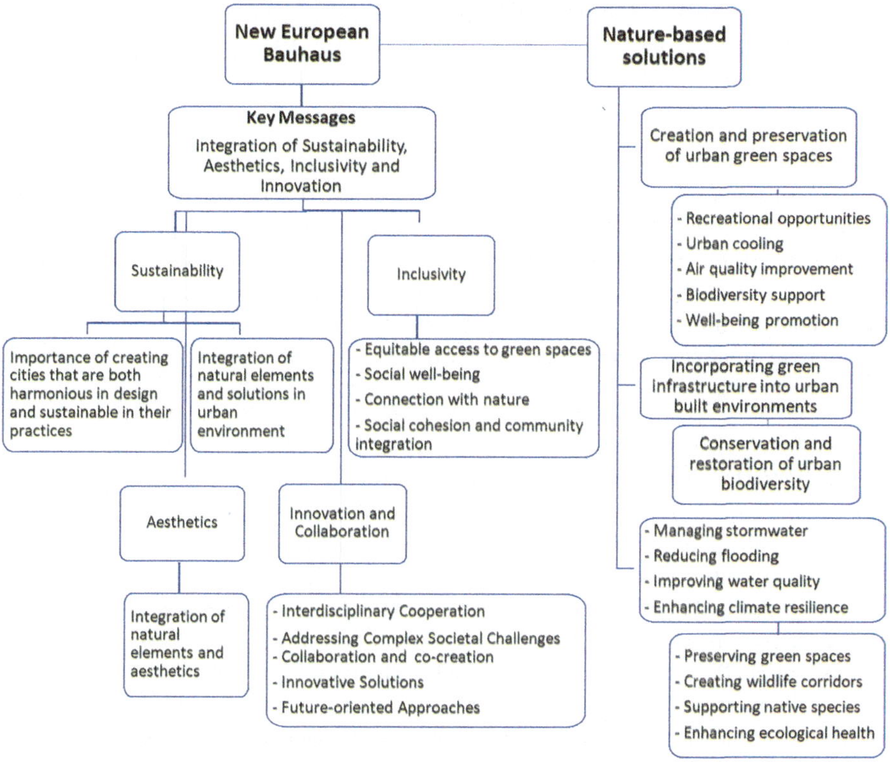

Fig. 4 Core elements of nature-based solutions and the New European Bauhaus

the risk of water pollution, and enhance the overall resilience of cities in the face of climate change. (3) *Conservation and restoration of urban biodiversity*: By preserving existing green spaces and creating wildlife corridors and rewilding, cities can support the survival of native species and promote ecological balance. Biodiversity in cities not only contributes to the overall health of urban ecosystems but also enhances the aesthetic value and cultural identity of the city.

The implementation of NBS and the integration of greenery into cities require a holistic approach that involves collaboration between various stakeholders, including urban planners, policymakers, architects, and community members. It is crucial to consider the specific context, climate, and needs of each city to develop tailored solutions that maximize the benefits of nature-based approaches [24].

In conclusion, NBS plays a vital role in creating beautiful, resilient, and healthy cities. They offer multiple advantages, including improved air and water quality, reduced urban heat, enhanced biodiversity, and increased well-being for urban dwellers. By embracing NBS approaches, cities can not only mitigate the environmental challenges they face but also create more livable and sustainable urban environments for both current and future generations.

4 Colored Solar Energy Technologies with Storage Flexibility for Zero GHG Emissions and Sustainable Cities

Solar energy technologies for the built and urban environment include the solar thermal systems (passive and active), photovoltaics (PV), and hybrid PV/thermal systems. The small dimensions of PV and the low-energy conversion make the systems ideal for their flexible integration into the urban environment either on the building as rooftop, façade or glazing application, landscape integration and agrivoltaics within or outside the city boundary, and PV integration in blue infrastructure as floated components over urban water bodies.

4.1 Rooftop Photovoltaics

The annual installation of rooftop solar PV (RTPV) has increased remarkably in the last years from 9.2 GW in 2010 to 76 GW in 2021, reaching a cumulative of more than 320 GW globally, with the number of RTPV household installations estimated to increase from 25 million today to more than 100 million by 2030 [25]. A high number of RTPV installations is expected in the coming years due to the very low cost of PV modules and the transition to clean energy in urban transformation. This rise has been initially driven by a few countries like Germany and Italy, but several other countries have set ambitious plans for sufficient electricity coverage by RTPVs [25]. This direction will be accelerated by the dramatical PV cost drop to more than 80% within the last decade, the simplicity of installation, and the flexibility in electricity generation matching the demand curve by the introduction of financial incentives like net metering, smart microgrids, and storage capabilities of durable batteries. In this context, the RTPVs are central in REPowerEU Action Plan and are expected to be mandatory for more than 250 m^2 roof areas by 2026 while the installation of RTPVs will be needed in all new residential buildings by 2029 to reach 600 GW by 2030.

A review assessment of practice city applications in Europe spanning different climates from cold north to temperate central and warm south shows that RTPVs are effective for coverage of building energy needs at all these zones.

4.2 BIPV and Storage Flexibility for Carbon-Free Urban Transformation

Photovoltaics can be integrated into almost all the solar-irradiated surfaces of the building while recent research efforts are also conducted for indoor applications with daylight conversion at the milliwatt scale. A thorough analysis performed recently by [17] showed that the prioritization of the integration can be aligned to the local climatic conditions with preferential PV façade integration in the cold and temperate climate of North and Central Europe and PV shading and PV glazing for the warm and hot climate conditions of South Europe, respectively [17]. For the BIPV to be fully utilized toward the carbon-free urban transformation, there is an urgent need to develop colored PV technology to expand the aesthetics limitations and the visually pleasant choices of the BIPV integration options and change the attitude of actors opposing PV utilization in the urban environment. BIPV aesthetics is among the most desired selection criteria for customers [26], and urban transformation with clean energy can be successful with the availability of colored BIPV choices for the architects and urban planners to combine with NBS and (re)create the beautiful landscapes of our cities. In this context, there are many available solutions by either colored photovoltaics such as organic cells and perovskites or additives for conventional modules such as antireflection coating and dielectric materials with adjusted thickness and structure for a variety of different colors such as yellow, blue, green, pink, and orange [26, 27].

For the substitution of the use of fossil fuels for space heating in temperate climate with solar energy, there are two main solutions that can be applied separately or combined. In the first, heat pumps driven by BIPV can be used to supply low-temperature heat to well-insulated buildings and neighborhoods. Second, district heating and urban networks can be designed as separate flexible systems or combined with heat pumps (and cover also older buildings or vulnerable people) with the energy of higher temperature provided by a variety of renewable sources such as PV and solar heating utility farms, fuel cells such as green hydrogen, surplus industrial heat, energy from waste treatment plants, and complementary renewables such as geothermal energy. In both solutions, heat storage through thermochemical reactions, sensible and phase changes [28], as well as NBS as blue infrastructure and pumped hydrostorage are important especially with the use of PV and should be integrated as the intermediate between the heat generation curve and the demand profile at different scales. Hydrogen generation can be considered as the storage option for excess renewable energy, which could be used as surplus energy, or carried for use in building or hydrogen cars. Therefore, stationary electrolyzers and tanks can be integrated within or outside the city boundary for hydrogen generation and storage, respectively. Due to the complexity of technological systems and for a successful urban transformation, cities will urgently need to proactively investigate the available renewable potentials and optimize the solutions at the different scales of buildings, neighborhoods, and districts after intense discussion with their citizens.

To facilitate the implementation of effective urban solutions in the carbon-free transformation, urban mesoscale atmospheric modeling based on the Weather Research and Forecasting (WRF) model [29] with the coupled urban modules [30] can be applied to assessing the sensitivity of the urban temperature and energy consumption upon land use, anthropogenic heat, green roofs, as well as photovoltaic panels modifications [31, 32]. With physically realistic simulations and atmospheric reconstruction of chosen meteorological events based on detailed urban mapping [32], and high horizontal resolution (0.5–1 km), strategies for better urban planning in view of heat, thermal comfort, and energy consumption considerations can be revealed with limited uncertainty.

4.3 Social Approaches to Accelerate the Carbon-Free Urban Transformation

As carbon-free urban transformation is complex, an emphasis on the integration of social, economic, and technological dimensions is necessary and inclusive and holistic approaches to energy transition and carbon-free urban transformation. Among these, public awareness and engagement, stakeholder collaboration, citizen participation and empowerment, behavioral change and sustainable lifestyles, social innovation, entrepreneurship, capacity building, and education can be considered. With a plethora of available tools as distributed energy cooperatives, integrated energy districts, localized energy hubs, circular economy districts, sharing economy platforms, and energy-positive communities, the appropriate selection and adaptation of these models are pivotal for addressing specific contextual factors, meeting societal needs, and fostering sustainable and resilient urban environments for all. Without investigating deeply the potential social consequences of the urban decarbonization processes, they might be unjust and fail.

5 Conclusions

By integrating and operationalizing NBS, technological solutions based on renewables such as solar energy, and social approaches, cities can harness the collective efforts of society and accelerate the carbon-free urban transformation.

Acknowledgment L.F.C. thanks the Catalan Government for the QA given to GREiA (2021 SGR 01615).

References

1. McKinsey Global Institute, The net-zero transition, What it would cost, what it could bring, Jan 2022.
2. United Nations Environment Programme, Cities and Climate change, https://www.unep.org/explore-topics/resource-efficiency/what-we-do/cities/cities-and-climate-change, last accessed 2023/07/2.
3. Lwasa, S., Seto, et al. Urban systems and other settlements. In IPCC: Climate Change : Mitigation of Climate Change. Contribution of WGr III to the 6th AR of IPCC [P.R. Shukla et al. (eds.)]. Cambridge University Press, Cambridge, UK and New York, NY, USA.(2022).

4. Huovila, A. et al. Carbon-neutral cities: Critical review of theory and practice. Journal of Cleaner Production, 341, 130912 (2022).
5. Report no. 05/2022, Air quality in Europe 2022 HTML – TH-AL-22-011-EN-Q – ISBN 978-92-9480-515-7 – ISSN 1977-8449 – https://doi.org/10.2800/488115.
6. European Environment Agency, EMEP/EEA air pollutant emission inventory guidebook (2016).
7. Khreis, H. et al. Urban policy interventions to reduce traffic-related emissions and air pollution: A systematic evidence map. Environment International, 172, 107805 (2023).
8. Breetz, H.L. et al. Small towns with big plans: Municipal adoption of 100% renewable electricity policies. Energy Research & Social Science 90, 102664 (2022).
9. Wang, D., et al. Neighbouring green space and mortality in community-dwelling elderly Hong Kong Chinese: a cohort study. BMJ Open. 7(7), e015794 (2017).
10. Connolly, R., et al. The association of green space, tree canopy and parks with life expectancy in neighborhoods of Los Angeles. Environment International, 173, 107785 (2023).
11. Rojas-Rueda, D. et al. Green spaces and mortality: a systematic review and meta-analysis of cohort studies. Lancet Planet. Health, 3, e469–e477 (2019).
12. European Commission 100 Climate-Neutral and Smart Cities by 2030 https://ec.europa.eu/info/news/commission-invites-cities-express-their-interest-become-part-european-mission-100-climate-neutral-and-smart-cities-2030-2021-nov-25_en (2022).
13. Vanhuyse, F., Piseddu, T., Jokiaho, J. Climate neutral cities in Sweden: True commitment or hollow statements?, Cities, 137, 104267 (2023).
14. European Commission, Info kit for cities interested in participating in the call for expression of interest (EOI) V2.0, https://research-and-innovation.ec.europa.eu/system/files/2021-11/ec_rtd_eu-mission-climate-neutral-cities-infokit.pdf (2021).
15. Greenhouse Gas Protocol. Available online: https://ghgprotocol.org, last accessed on 28 June 2023.
16. van den Dobbelsteen, A.The Regenerative city: positive opportunities of coupling urban energy transition with added values to people and environment. In: Roggema, R. (eds) TransFEWmation: Towards Design-led Food-Energy-Water Systems for Future Urbanization. Contemporary Urban Design Thinking. pp. 235–252 Springer Nature, Switzerland (2021).
17. Skandalos, N., et al. Building PV integration according to regional climate conditions: BIPV regional adaptability extending Köppen-Geiger climate classification against urban and climate-related temperature increases. Renewable Sustainable Energy Reviews, 169, 112950 (2022).
18. C40 cities and ARUP, Green and thriving neighborhoods guidebook, A pathway to net zero, featuring the '15-minute city'. (2021).
19. Liu, H.-Y., Jay, M., Chen, X. The role of nature-based solutions for improving environmental quality, health and well-being. Sustainability, 13, 10950 (2021).
20. Churkina, G., et al. The effects of urbanization and climate change on urban temperatures: A systematic review. Urban Climate, 18, 66–82 (2016).
21. World Health Organization. Urban Green Spaces and Health: A review of evidence. (2016).
22. Lundholm, J. T., et al. Urban Ecosystems: Ecological Principles for the Built Environment. Cambridge University Press. (2015).
23. European Commission. The New European Bauhaus – The vision. (2021).
24. Liu, H.-Y., et al. Integrating solar energy and nature-based solutions for climate-neutral urban environments. Solar, 3, 382–415 (2023).
25. IEA. Technology and innovation pathways for zero-carbon-ready buildings by 2030, IEA, Paris https://www.iea.org/reports/technology-and-innovation-pathways-for-zero-carbon-ready-buildings-by-2030, License: CC BY 4.0) (2022).
26. Basher, M.K. et al. Aesthetically appealing building integrated photovoltaic systems for net-zero energy buildings. Current status, challenges, and future developments—A review. Buildings. 13(4),863 (2023).

27. Xu, Z. et al. H. Tunable and angle-insensitive structural coloring of solar cell modules for high performance building-integrated photovoltaic application. Solar Energy Materials and Solar Cells 247, 111952 (2022).
28. Hesaraki, A., Holmberg, S., Haghighat, F. Seasonal thermal energy storage with heat pumps and low temperatures in building projects – A comparative review. Renewable Sustainable Energy Reviews, 43, 1199–1213 (2015).
29. Skamarock, W.C. et al. NCAR technical note, a description of the advanced research WRF Model Version 4.3. (2021).
30. Chen, F. et al. The integrated WRF/Urban modelling system: development, evaluation, and applications to urban environmental problems. International J Climatology 31, 273 (2011).
31. Zonato, A. et al. Exploring the effects of rooftop mitigation strategies on urban temperatures and energy consumption. Journal Geophysical Research: Atmospheres 126(21) (2021).
32. Avisar, D. et al. High resolution WRF simulations for the Tel-Aviv metropolitan area reveal the urban fingerprint in the sea-breeze hodograph. Journal Geophysical Research: Atmospheres 126, 1 (2021).

South Africa's Solar Rebates: Hope or Hate for Residential Customers Within eThekwini Municipality

Leshan Moodliar and Innocent Ewean Davidson

1 Introduction

1.1 Load Shedding

Load shedding is a control measure implemented by the System Operator in South Africa to maintain a balanced supply and demand of electricity while maintaining the system frequency at 50 Hz. This method has existed since 2008, but incidents of load-shedding have become increasingly common in recent times. EThekwini Municipality implements load shedding in eight stages, rotating among 24 pre-identified blocks. The interruption frequencies and average weekly times without electricity when load shedding is implemented are outlined in Table 1.

Load shedding has been a persistent issue in South Africa, with stages 4–6 being frequently implemented since the beginning of 2023. Should load shedding stages 4-6 continue, then residents in eThekwini would experience power outages ranging between 36 and 50 h per week.

1.2 The Status of Electricity Supply in South Africa

Load shedding in South Africa is necessary as there is inadequate generation capacity to meet the projected demand. Based on the data presented in Fig. 1a, the operating reserve for the second week of April 2023 indicates a reserve margin that falls, on average, 11% short of the daily demand.

L. Moodliar (✉) · I. E. Davidson
Cape Peninsula University of Technology, Cape Town, South Africa
e-mail: 240073800@mycput.ac.za

Z. Peng (ed.), *Trends and Technological Challenges in Green Energy*, Green Energy and Technology, https://doi.org/10.1007/978-3-031-55579-4_3

Table 1 Load-shedding interruption frequency per stage per week [1]

Load-shedding stage	1	2	3	4	5	6	7	8
Average number of load-shedding events per week	5	10	12	18	21	25	30	31
Number of hours without electricity per week (h)	10	20	24	36	42	50	60	62
% Time without electricity in the week	5.95	11.9	14.2	21.4	25.0	29.7	35.7	36.9

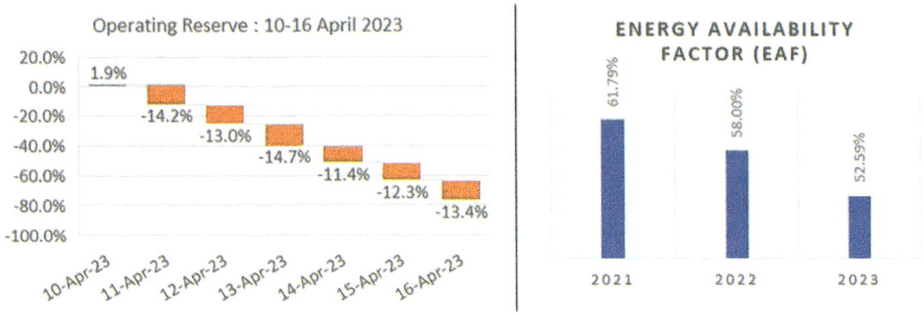

Fig. 1 (**a**) National generation performance. (Adapted from [2]); (**b**) energy availability factor (EAF). (Adapted from [2, 3])

The energy availability factor (EAF) is a performance indicator used to measure the reliability and availability of power stations. As depicted in Fig. 1b, the EAF for Eskom power plants has decreased from 61.79% in 2021 to 58.00% in 2022 and 52.59% in 2023. It is therefore imperative to implement significant interventions to improve plant performance and increase capacity. The South African government has developed a national plan to address this situation, which includes investing in diversified generation, upgrading and maintaining existing infrastructure, promoting demand-side energy-savings initiatives, and implementing structural reforms within the electricity sector.

1.3 National Response to Load Shedding in South Africa

The South African government has developed the Energy Action Plan (EAP) based on five strategic pillars to address the energy crisis and end load shedding. The plan was introduced in 2022 and is overseen by the National Energy Crisis Committee (NECOM). Figure 2 provides an overview of the interventions outlined in the EAP.

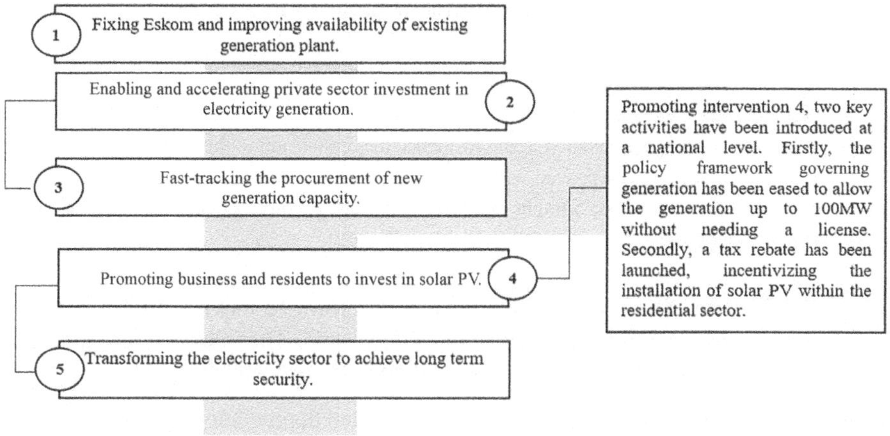

Fig. 2 Load shedding mitigation interventions as outlined in the EAP [4]

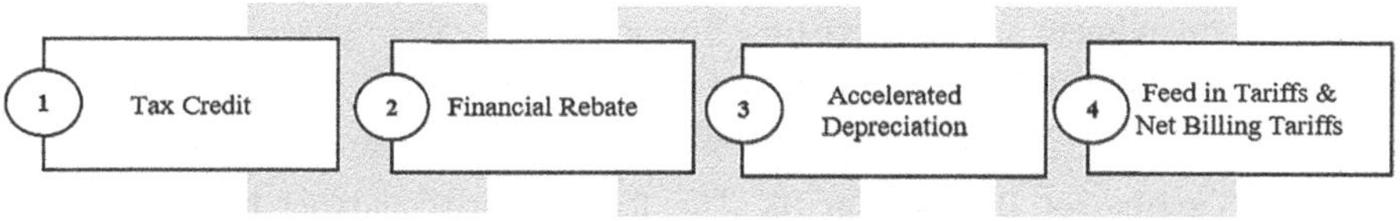

Fig. 3 Incentives aimed at improving renewable energy adoption rates

1.4 Incentives for Generation

Solar-generated electricity is quickly complementing grid-supplied electricity. In eThekwini, solar PV installations are increasing, lowering grid electricity use. However, the high cost of technology and regulatory uncertainty in South Africa have historically hindered the widespread rollout of solar PV. Internationally, incentives were the main drivers in promoting renewable energy. With the help of incentives, solar PV becomes more affordable and adoption rates increase. As illustrated in Fig.3, incentives for solar installations are packaged differently. However, they all have the effect of lowering the overall cost of ownership [5–7] (Fig. 3).

Tax Credit Many countries offer a tax credit incentive for those who invest in solar energy [6]. This means that customers can receive a portion of their solar investment back through their annual tax assessment process [7]. The specific amount that can be claimed varies by location. In South Africa, up to 25% of the solar panel cost, limited to R 15,000, may be claimed, depending on the criteria outlined in Table 2.

Table 2 Qualifying criteria for the solar PV rebate as introduced by SARS [8]

No.	Criteria
1	Only applicable to new solar panels with a minimum capacity of 275 W each
2	Portable solar panels do not qualify for the incentive
3	Solar panels are to be installed at the residence and connected to the main supply
4	The installation must be proved with a certificate of compliance
5	New solar panels must be brought into service for the first time between March 1, 2023, to February 29, 2024

Financial Rebate The financial rebate takes the form of making an upfront payment to the customer for installing the solar PV system. The impact is similar to the tax rebate; however, rebates are not done via the tax mechanism.

Accelerated Depreciation Allowing for accelerated depreciation is another form of incentivizing the installation of solar PV [9]. This mechanism is largely utilized within the business sector. Energy-generating assets are subject to a faster rate of depreciation than traditional assets, resulting in accelerated tax savings. In 2016, the South African Revenue Service (SARS) amended Section 12b of the Income Tax Act. The amendment saw the existing 3-year (year 1: 50%; year 2: 30%; year 3: 20%) accelerated depreciation allowance on renewable energy shift to a single year at the full cost, that is, 100% depreciation in year 1 [10].

Feed in Tariffs/Net Billing Tariffs The feed in tariff and net billing are also incentives; however, they are not upfront, and the benefit is reaped throughout the lifespan of the project as energy is generated [11]. The tariffs are usually guaranteed at a specified rate and for a specific time period. With the guaranteed rate over the predefined time period, securing funding for the system becomes easier. The feed in tariff and the net billing tariff help reduce the SPBP and strengthen the investment return of the solar PV project. Feed in tariffs are extremely popular and have been used to accelerate the energy transition in many countries, including Germany and Spain [12, 13]. The eThekwini Municipality has introduced a net billing tariff aimed at the residential sector where the buyback rate for energy generated into the grid is R 1.08/kWh (Table 3). As illustrated in Table 3, incentive mechanisms vary per country. The extent and variety of the incentive mechanism are influenced by country-specific policy.

Incentives have been crucial in driving the energy transition in many countries. The level of incentives will vary depending on the country's circumstances. The higher the incentive, the lower the overall price of the renewable energy system and generally the tendency to adopt the technology increases. A higher adoption rate of renewable energy in South Africa could help alleviate the constrained grid and reduce greenhouse gas emissions as reliance on fossil fuels is reduced.

Table 3 Incentive mechanism and country of implementation [14]

Subsidy type→ Country↓	Feed-in tariff	Net metering	Tax-related subsidy	Funding – interest related	Other
Germany	•	•	•	•	•
Spain	•			•	•
France	•			•	•
China	•	•	•		•
India	•	•	•		•
Malaysia	•	•			
Canada	•	•			•

2 Literature Review

Incentives and rebates have been identified as important factors for promoting renewable energy. A range of studies have examined the efficacy of incentives in promoting the installation of renewable energy [15].

A study by [16] evaluated the various factors that promote the uptake of solar PV. These factors include environmental, personal preferences, and economic. It was concluded that households place great emphasis on the costs versus benefit principles when investing in solar PV. Hence, the provision of financial incentives is instrumental in influencing the decision to invest in solar PV. In a research study labeled "When the carrot goes bad: the effect of solar rebate uncertainty" by [17], it is confirmed that incentive policies including tax credits are important factors encouraging solar adoption. The research further indicated the importance of stability within rebate policies as it was concluded that policy uncertainty would lower adoption rates. The research by [18] highlighted an important aspect of information dissemination to the customer base as it positively affected solar PV adoption rates. It is, therefore, important to ensure that available incentives are appropriately communicated to the sector. The authors of [19] have investigated the factors promoting solar PV uptake in Malta; among other factors, they agreed that financial motives and constraints are key drivers. A study by [20] in Finland concluded that customers in good financial circumstances would promote PV uptake irrespective of the economic gains. It was cited that own PV generation brought pleasures to the adopters. Research by [21] highlights that the development of the energy market in the United States was hinged on a range of policies and incentives. Each policy and incentive, however, has a different impact on adoption rates, and not all incentives and policies result in equal adoption rates.

Based on the policy-stacking theory, a range of policies and incentives introduced at appropriate time intervals based on local circumstances is the optimum approach to maximizing the uptake rate of solar PV [22]. The policy-stacking theory for distributed generation is based on three phases. The initial phase is market preparation, thereafter market creation, and, finally, market expansion. Subsidies and incentives play a key role in market creation and expansion [23].

All the reviewed studies indicate that rebates are effective in promoting the adoption of solar PV. However, the effectiveness of the rebate would depend on various factors, including market structure, maturity of the market, the level of rebate applied, and the existing costs of solar PV systems weighted against the local grid-supplied electricity rates.

3 Research Methodology and Outputs

The study's aim is to evaluate the combined impact of the solar rebate as introduced by SARS and the residential net billing tariff as introduced by eThekwini Municipality on the viability of residential solar PV projects. According to Fig. 4, the methodology includes gathering a range of input data and subjecting such data to techno-economic analysis through scenario modeling as per Table 4. The analysis was carried out utilizing the Hybrid Optimization of Multiple Energy Resources (HOMER) modeling software. The range of outputs included the financial savings,

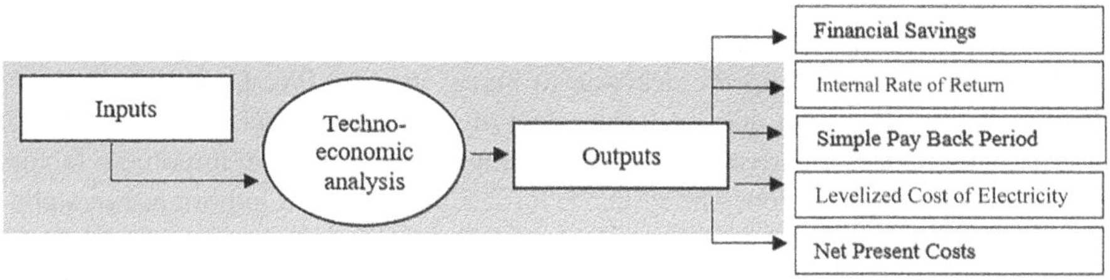

Fig. 4 Techno-economic modeling process flow

Table 4 Scenario modeling

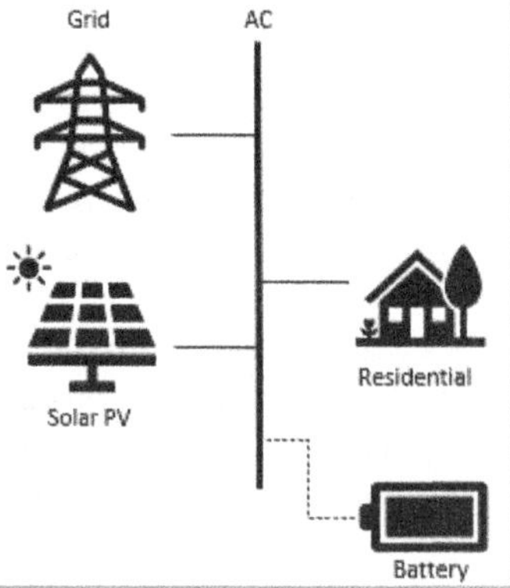

Scenario modeling
Base case: grid-only supply *Scenario 1*: 5 kVA solar PV + 5 kWh BESS interconnected with the grid *Scenario 2*: 5 kVA solar PV + 5 kWh BESS + tax incentive + net billing tariff, interconnected with the grid

internal rate of return (IRR), simple payback period (SPBP), levelized cost of electricity (LCOE), and the net present cost (NPC). These outputs were used to evaluate the project feasibility under the various scenarios.

4 Modeling Analysis and Scenario Setting (Table 5)

To effectively evaluate the impact of the solar rebate and the net billing incentive, the scenarios as per Table 4 have been modeled. The key input data applicable for modelling are outlined in Table 5.

5 Discussion of Results (Table 6)

The results for scenarios 1 and 2 are presented in Table 6. Without any subsidy or incentives per scenario 1, the solar PV and BESS project is able to save the customer R 15,129 per annum with an initial investment of R 112,500. The project produces an IRR of 12.80% with a SPBP of 6.18 years. The LCOE is improved by 0.67 c/kWh compared to the base case. Applying the tax incentive and the municipal net billing tariff as per scenario 2 reduces the yearly savings by R 159.00; however, with a reduced upfront payment of R 97,500, the IRR has improved to 15.50% while the SPBP reduced to 5.57 years. The LCOE has improved to R 2.80/kWh while the NPC reduced by R 11,755. With improved investment criteria, the installation of solar PV and BESS becomes more lucrative.

While the subsidy is able to improve the investment metrics of the project and will encourage a greater uptake of solar PV, it does result in less electricity being bought

Table 5 Input data applicable for scenario modeling

Residential loading	*Battery energy storage*	*Solar PV system*	*Electricity grid (residential net billing tariff)*	*General*
Loading: 35 kWh/day	Battery size: 5 kWh Battery cost: R 7500/kWh Battery replacement year: 10 years	PV System size: 5 kVA PV system installed costs: R 15,000/kVA Replacement inverter costs: R 5000/kVA Replacement year: 10 years	Municipal purchase rate: R 2.58/kWh Municipal net billing incentive: R 1.08/kWh Municipal network access charge: R 22/kVA (based on inverter size)	Modelling period: 10 years Nominal discount rate: 12% SARS solar rebate: R 15,000

Table 6 Key results for modeling scenarios 1 and 2.

Modeling output	Unit	*Base case (grid only)*	*Scenario 1*: 5 kVA solar PV + 5 kWh battery energy storage	*Scenario 2*: 5 kVA solar PV + 5 kWh battery energy storage + net billing tariff as applied by eThekwini Municipality + tax incentive as applied by SARS
Capital investment	(R)	–	112,500	97,500
Financial savings	(R)/ year	–	15,129	14,970
Internal rate of return (IRR)	%	–	12.80	15.50
Simple payback (SPBP)	Year	–	6.18	5.57
Levelized cost of electricity (LCOE)	R/ kWh	3.59	2.92	2.80
Net present cost (NPC)	(R)	326,829	285,656	273,901

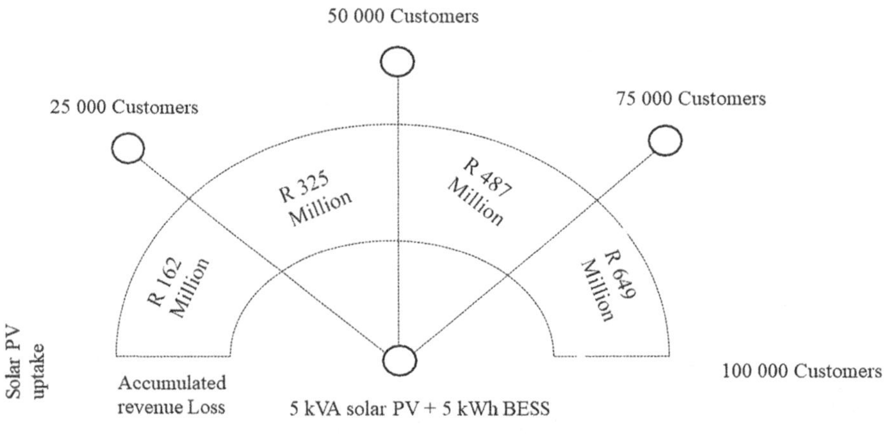

Fig. 5 Municipal revenue loss versus solar PV uptake

from the municipality. As a result, the municipality buys less however, the residential sector tariffs in eThekwini recover all of the network costs (100%) via the energy charge [24]. Therefore, each kWh not sold results in the fixed network costs not being recovered, resulting in a direct financial loss to the municipality. Figure 5 depicts the financial losses applicable to the municipality as solar PV and BESS installations increase. To counteract the loss and keep the municipality revenue neutral, the fixed NAC needs to be increased from R 22.00/kVA to R 108.00/ kVA. However, in doing so, the financial metrics of the project is severely

Table 7 Modeling results with the increased NAC

NAC per month	Financial savings	Internal rate of return (IRR)	Simple payback period (SPBP)	Levelized cost of electricity (LCOE)	Net present costs (NPC)
R/kVA	R/year	%	year	R/kWh	R
108.00	9,810	7.00	7.93	3.32	325,090

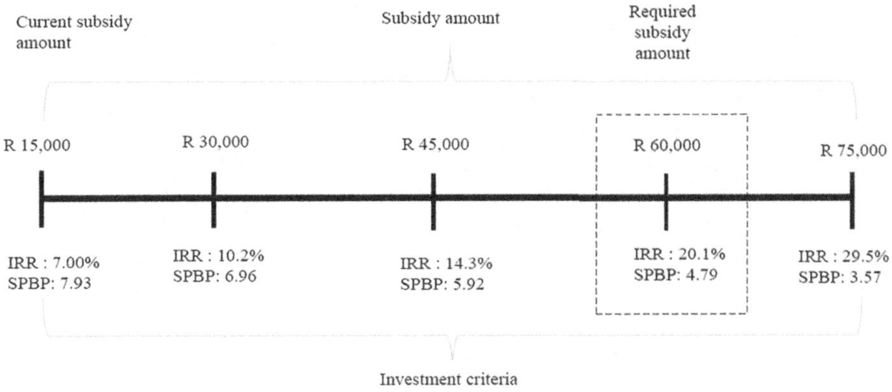

Fig. 6 Investment metrics versus subsidy provision

weakened. Table 7 indicates the key results per scenario 2, however, with at a fixed NAC of R 108.00/kVA per month.

To keep the municipality revenue neutral and improve the investment metrics of the project, the upfront subsidy needs to increase. Figure 6 highlights the improvement in the financial metrics as the subsidy is increased. An upfront subsidy of R 60,000 is required to keep the municipality revenue neutral and generate a project IRR greater than 20%, enabling a lucrative incentive to the residents of eThekwini to install solar PV and BESS systems. The upfront subsidy required is four times the subsidy currently being offered by SARS.

6 Conclusion

Solar rebates have proven to be an effective tool in promoting the use of solar energy worldwide, although their structures varies depending on the country. Recently, the SARS introduced a solar rebate program aimed at the residential sector that provides an upfront subsidy of up to R 15,000 for installing solar panels. In addition, the eThekwini Municipality offers a residential net billing tariff that provides for a buyback rate for energy generated into the grid, further incentivizing the use of solar energy. These incentives effectively improve the investment metrics of a 5 kVA solar PV installation coupled with 5 kWh of BESS. With the upfront subsidy and the net billing tariff, the IRR increased from 12.80% to 15.50% while the SPBP decreased from 6.18 years to 5.57 years. The overall NPC also improved by R 11,755. Investing in solar energy has become marginally more feasible within eThekwini municipality due to these current incentives.

However, the current residential tariff structure lacks fixed charges for network cost recovery, resulting in revenue losses for the eThekwini Municipality due to solar installations. If one hundred thousand residential customers install solar PV with BESS, the municipality could potentially lose R 649,000,000. To mitigate against these losses, the NAC within the net billing tariff must increase from R 22/kVA to R 108/kVA per month. While this solution solves the municipal loss issue, it weakens the investment criteria of the solar projects as the IRR drops from 15.50% to 7.00% and the SPBP increases from 5.57 to 7.93 years. Alternatively, increasing the upfront subsidy from R 15,000 to R 60,000 would enable the municipality to levy the higher NAC (i.e., R 108.00/kVA), eliminating their losses whilst also boosting the project IRR to 20.1% and reducing the SPBP to 4.79 years. As the priority to shift to cleaner energy sources intensifies, solar incentives will continue to play a crucial role in encouraging renewable energy adoption.

Acknowledgment The researchers express their gratitude to the Electricity Unit of eThekwini Municipality for their valuable assistance and provision of data during the study.

References

1. EThekwini_Electricity_Unit. Load Shedding Schedule. (2019). Accessed: January 2020. [Online]. Available: http://www.durban.gov.za/City_Services/electricity/Load_Shedding/Pages/default.aspx
2. [2] Eskom, "Weekly System Status Report – 2023 Week 15 (10/04/2023 – 16/04/2023)," 2023. Accessed: 20 April 2023 [Online]. Available: https://www.eskom.co.za/wpcontent/uploads/2023/04/Weekly_System_Status_Report_2023_w15.pdf
3. Eskom, "Weekly System Status Report – 2022 Week 1 (03/01/2022 – 09/01/2022)," 2022. Accessed: 13 April 2023.[Online]. Available: https://www.eskom.co.za/wp.content/uploads/2022/01/Weekly_System_Status_Report_2022_w1.pdf
4. (2022). Confronting the energy crisis: An action plan to end load shedding. [Online] Available: https://www.gov.za/documents/confronting-energy-crisis-action-plan-end-load-shedding-25-jul-2022-0000
5. Y. Heng, C.-L. Lu, L. Yu, and Z. Gao, "The heterogeneous preferences for solar energy policies among US households," Energy Policy, vol. 137, p. 111187, 2020.
6. M. Bhattacharya, S. R. Paramati, I. Ozturk, and S. Bhattacharya, "The effect of renewable energy consumption on economic growth: Evidence from top 38 countries," Applied energy, vol. 162, pp. 733–741, 2016.
7. C. Crago and I. Chernyakhovskiy, "Solar PV technology adoption in the United States: an empirical investigation of state policy effectiveness," 2014.
8. (2023) Solar panel.tax incentive for individuals [Online]. Available: https://www.treasury.gov.za/documents/National%20Budget/2023/2023%20Budget%20FAQs%20-%20Solar%20Panel%20Tax%20Incentive.pdf
9. F. Z. El-Karmi and N. M. Abu-Shikhah, "The role of financial incentives in promoting renewable energy in Jordan," Renewable Energy, vol. 57, pp. 620–625, 2013.

10. (2020). Tax Guide for Small Businesses. [Online] Available: https://www.sars.gov.za/wp-content/uploads/Ops/Guides/LAPD-Gen-G09-Tax-Guide-for-Small-Businesses.pdf
11. Y. Yamamoto, "Feed-in tariffs combined with capital subsidies for promoting the adoption of residential photovoltaic systems," Energy Policy, vol. 111, pp. 312–320, 2017.
12. A. Poullikkas, G. Kourtis, and I. Hadjipaschalis, "A review of net metering mechanism for electricity renewable energy sources," International Journal of Energy and Environment (Print), vol. 4, 2013.
13. J. A. Lesser and X. Su, "Design of an economically efficient feed-in tariff structure for renewable energy development," Energy policy, vol. 36, no. 3, pp. 981–990, 2008.
14. U. Kılıç and B. Kekezoğlu, "A review of solar photovoltaic incentives and Policy: Selected countries and Turkey," Ain Shams Engineering Journal, vol. 13, no. 5, 2022, https://doi.org/10.1016/j.asej.2021.101669.
15. L. Zhao, Y. Zhang, M. Sadiq, V. M. Hieu, and T. Q. Ngo, "Testing green fiscal policies for green investment, innovation and green productivity amid the COVID-19 era," Economic Change and Restructuring, pp. 1–22, 2021.
16. A. Jacksohn, P. Grösche, K. Rehdanz, and C. Schröder, "Drivers of renewable technology adoption in the household sector," Energy Economics, vol. 81, pp. 216–226, 2019.
17. T. Tsvetanov, "When the carrot goes bad: The effect of solar rebate uncertainty," Energy Economics, vol. 81, pp. 886–898, 2019.
18. A. Palm and B. Lantz, "Information dissemination and residential solar PV adoption rates: The effect of an information campaign in Sweden," Energy Policy, vol. 142, p. 111540, 2020.
19. M. Briguglio and G. Formosa, "When households go solar: Determinants of uptake of a Photovoltaic Scheme and policy insights," Energy Policy, vol. 108, pp. 154–162, 2017.
20. S. Karjalainen and H. Ahvenniemi, "Pleasure is the profit-the adoption of solar PV systems by households in Finland," Renewable energy, vol. 133, pp. 44–52, 2019.
21. A. J. Ryan, F. Donou-Adonsou, and L. N. Calkins, "Subsidizing the sun: The impact of state policies on electricity generated from solar photovoltaic," Economic Analysis and Policy, vol. 63, pp. 1–10, 2019.
22. V. A. Krasko and E. Doris, "State distributed PV policies: Can low cost (to government) policies have a market impact?," Energy Policy, vol. 59, pp. 172–181, 2013.
23. E. Doris, "Policy building blocks: helping policymakers determine policy staging for the development of distributed PV markets," National Renewable Energy Lab.(NREL), Golden, CO (United States), 2012.
24. EThekwini_Municipality_:_Electricity_Unit. Tariff Booklet 2022_2023 [Online] Available: https://www.durban.gov.za/pages/government/documents?d=Service%20Tariffs/Electricity%20Tariffs

A Review of LED Driver Topologies and Control Methods for Energy-Efficient Smart Farming Application

M. L. Pay, J. Christensen, F. He, L. Roden, M. Rasel, L. Tyler, H. Ahmed, and M. Foo

1 Introduction

The rising global population is causing increased food demand all over the world. To keep up with this ever-increasing demand, there is an urgent need to increase food production substantially. However, the increasing severity and frequency of environmental stressors exacerbated by climate change, such as prolonged drought, excessive flooding, and unpredictable changes in temperature, are causing serious problems to the yield and quality of food through conventional farming practices. Such issues lead to uncertainties in food production. Additionally, increasing the agricultural intensity and/or land use for food production can significantly accelerate

M. L. Pay (✉) · J. Christensen
Institute for Future Transport and Cities, Coventry University, Coventry, UK
e-mail: paym@uni.coventry.ac.uk

F. He
Centre for Computational Science and Mathematical Modelling, Coventry University, Coventry, UK

L. Roden
Centre for Sport, Exercise and Life Sciences, School of Life Sciences, Coventry University, Coventry, UK

M. Rasel
Department of Genetics and Plant Breeding, Bangladesh Agricultural University, Mymensingh, Bangladesh

L. Tyler
Menter Môn Cyf, Llangefni, Gwynedd, UK

H. Ahmed
Nuclear AMRC Midlands, University of Sheffield, Derby, UK

M. Foo
School of Engineering, University of Warwick, Coventry, UK

Z. Peng (ed.), *Trends and Technological Challenges in Green Energy*, Green Energy and Technology, https://doi.org/10.1007/978-3-031-55579-4_4

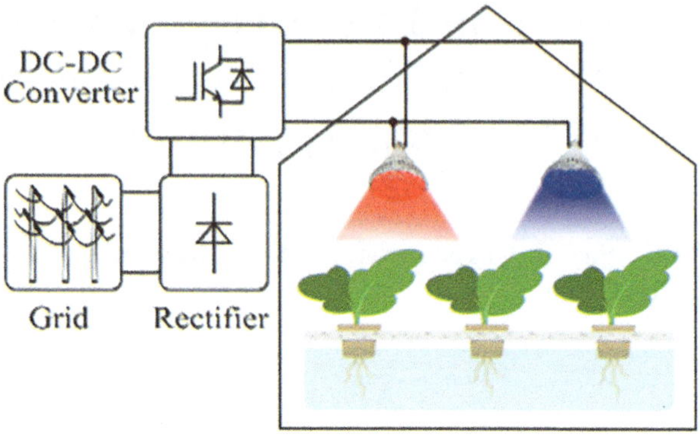

Fig. 1 An overview of the configuration of LED driver for light manipulation in smart farming

the negative impacts of climate change. In this context, smart (vertical) farming [1] where plants are grown in a controlled environment using artificial lights, and mostly using hydroponics where crops are fed through a nutrient solution, has recently gained popularity. It is possible to improve the efficiency of crop production in smart farming by manipulating various light qualities, such as intensity, photoperiod, and wavelength [2, 3]. In addition, smart farming can increase food production without increasing agricultural land use [4].

In smart farming, light-emitting diodes (LED) are a popular choice for plant growth as LEDs have low operational cost and do not produce as much heat as other light sources. Moreover, the availability of different LED colors allows specific combinations to facilitate manipulation of light properties according to user demands. LEDs require direct current (DC) power, whereas the grid provides alternating current (AC) power, necessitating the use of a rectifier and DC–DC converter as depicted in Fig. 1. This motivates the development of LED drivers that take AC input from the grid and provide DC output to LEDs. However, LEDs are classified as nonlinear loads. It is well known that nonlinear loads can cause harmonic distortion, thereby increasing the power loss and decreasing the efficiency. Improving the efficiency of LED drivers can play a big role not only in optimal manipulation of light properties but in reducing the overall energy consumption in smart farming. To improve the power quality issue caused by these nonlinear loads, power factor correction (PFC) circuits are widely used in the LED driver improvement [5]. In addition, this also plays a role in reducing carbon emissions and improving food production sustainability.

From the circuit viewpoint, grid-connected LED drivers consist of two stages: the rectification of the AC voltage and the DC–DC conversion. In most cases, passive diode-bridge (DB) rectifiers are used to reduce the cost and implementation complexity. However, active rectification using power semiconductor switches is also frequently considered [6]. Once the rectified voltage is obtained, the DC–DC power stage provides the required power conversion to supply constant LED driver output. This stage also includes the PFC operation to make the driver grid-friendly. Meanwhile, the control system, which generates the switching signals for the power semiconductor switches in LED drivers, also plays a fundamental role in improving their efficiency. The control system ensures the synchronous operation of the LED driver with the grid. It can reduce the total harmonic distortion (THD) and switching losses. Owing to the importance of control systems in regulating the LED driver output, this study aims to provide a review of the conventional circuit topologies and control methods used in LED drivers suitable for smart-farming applications. This review discusses various control methods, compares their strengths and limitations, and investigates some future research directions.

The rest of this chapter is organized as follows: Sect. 2 provides a short summary of the selected popular topologies of PFC circuits for potential application in LED drivers; Sect. 3 summarizes the control methods; and Sect. 4 provides conclusion and future research perspectives.

2 Power Factor Correction (PFC) Circuit Topologies

PFC converters are the most widely used driver circuit for LED applications. In the basic operation, the circuits convert AC power from the grid into pulsating DC through a diode-bridge (DB) rectifier. This rectified output is then passed through the DC–DC boost converter to maintain the desired output voltage. Most of the low-cost PFC converters use a transformer to reduce the input voltage, which allows the use of low-voltage power semiconductor devices and provides galvanic isolation.

Out of various topologies available in the literature, the boost PFC is the most commonly used topology [7–9]. A circuit diagram of the conventional boost PFC is provided in Fig. 2a. This circuit is very simple to construct and requires only a few components. In this circuit, a transformer is first used to lower the voltage, then, the boost operation increases the voltage to a desired level that is suitable for driving the LED lights. This decreases the output voltage ripple and ensures synchronous operation of input current with the grid voltage. Although conventional

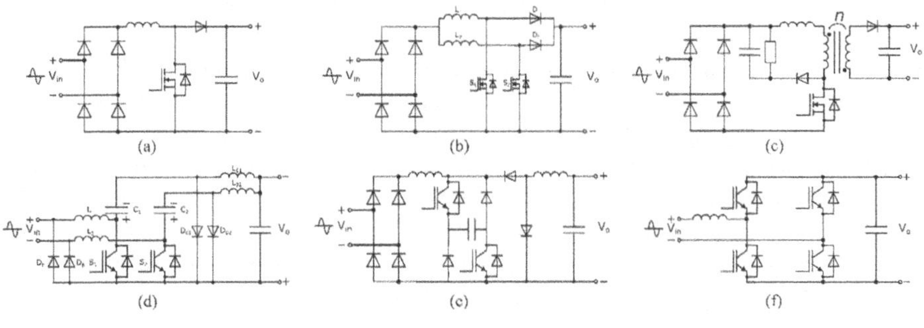

Fig. 2 Topologies of conventional PFC circuits suitable for LED driver application: (**a**) boost PFC converter, (**b**) interleaved boost PFC converter, (**c**) flyback PFC converter, (**d**) Ćuk PFC converter, (**e**) Sheppard–Taylor PFC converter, and (**f**) Totem–Pole PFC converter

single-boost PFC can handle higher power, interleaved boost PFC circuits [10, 11] (*cf.* Fig. 2b) are preferred for high-power applications. In the interleaved boost PFC, multiple boost PFC circuits are interleaved with phase shifts to improve the power-handling capability. This can also reduce the input current ripple and increase the efficiency.

The flyback PFC converter [12] (*cf.* Fig. 2c) is another popular topology. This circuit consists of a transformer used as an energy storage and provides isolation. This topology can provide multiple outputs and works effectively as a buck–boost converter. The bridgeless Ćuk PFC converter [13] (*cf.* Fig. 2(d)) uses capacitors as the energy storage and can provide limited isolation. The inductors in this topology can be implemented as separate or coupled inductors. Moreover, the inductors can easily be sourced locally, which reduces the circuit development time unlike some other topologies. The Sheppard–Taylor converter (STC) [14, 15] (*cf.* Fig. 2e) is similar to the Ćuk PFC converter as they both use two inductors. Unlike the Ćuk PFC converter, which suffers from control detuning issues at a very low value of the input voltage, the STC overcomes this issue by using two synchronized switches.

Another available topology is the Totem pole PFC converter [16, 17] (*cf.* Fig. 2f). In this bridgeless PFC circuit, rectification and boost operation are achieved by using four power semiconductor switches. This circuit can reduce the switching loss, leading to the efficient operation of the LED driver. There are many other circuits described in the literature. Most of them are extended versions of the six common topologies that are summarized in this section. As such, the extended topologies are avoided here for the sake of brevity. Interested readers may consult [6] and the references therein for other LED driver topologies.

3 Power Factor Correction (PFC) Circuit Control Methods

As discussed previously, LEDs are typically powered through PFC converters. These converters draw nonlinear currents, which results in poor power quality issues at the grid. Control techniques play an important role in reducing the THD and power loss in the power conversion process. By reducing the loss and increasing the efficiency, control techniques can make the overall system economically competitive. An overview of the PFC converter-based LED driver with control system is given in Fig. 3. Due to this important role, numerous control methods for PFC converters are described in the literature. Conventionally, PFC converters are controlled through a two-loop architecture (*cf.* Fig. 3). The slower outer loop is used for stabilizing the output voltage to a desired fixed value. The faster inner-loop current controller generates the switching signals for the power semiconductor switches to ensure unity power factor operation. In this kind of architecture, the grid, load voltage sensors, and an input current sensor are required. A current sensorless control method is also available [18], where this signal is estimated not measured.

The conventional proportional integral (PI) controller is a popular choice for the voltage loop. The PI controller is characterized by two gains, namely, the

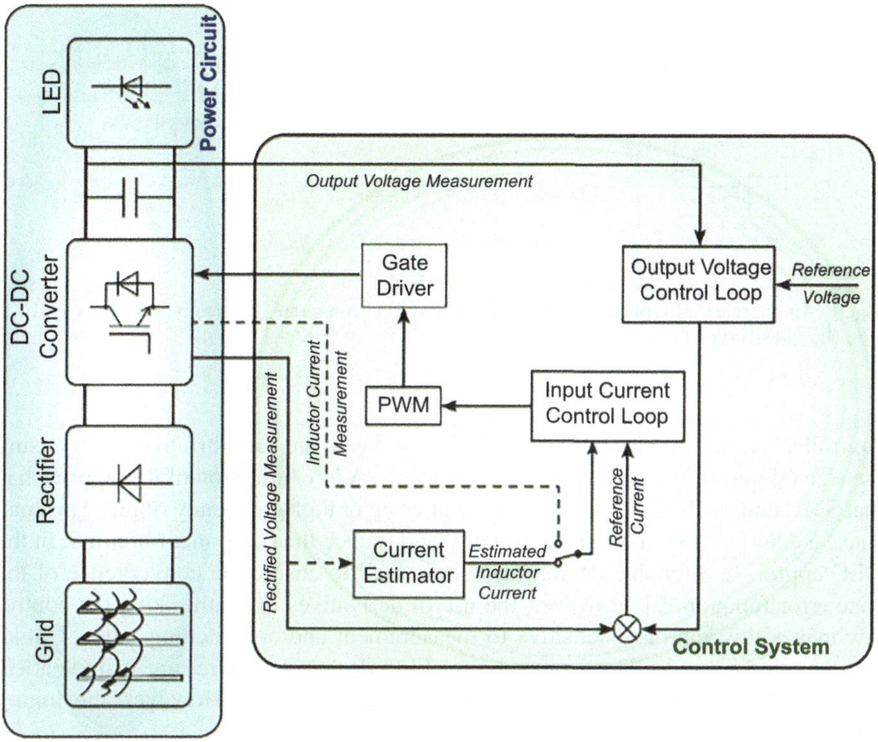

Fig. 3 An overview of PFC-based LED driver control system. The actual current signal (dashed line) can be obtained either by using estimation technique or through current sensor

proportional and integral gains. These gains can be tuned either through trial-and-error and/or using a model-based approach. As the outer loop deals with the slower voltage dynamics, most of the advanced control methods modify the current controller to achieve the desired performance. According to Lopez et al. [19], current controllers for PFC converters are typically classified into four types as shown in Fig. 4. In type I, PFC converters typically operate in discontinuous conduction mode or in the boundary between continuous and discontinuous conduction mode. Nonlinear techniques, such as hysteresis, sliding mode control (SMC), high-order SMC (HOSMC), dynamic evolution control (DEC), and fuzzy logic controller (FLC), are applied in type II control methods.

Nonlinear control methods in type II rely on using the sign of the tracking error in generating the gating pulses. In the hysteresis control, the tracking error is passed through a relay block without using any tuning gain [20, 21]. Although this is simple to implement, it may cause a significant ripple in the current. SMC [7, 11] and HOSMC [22] typically operate by defining a sliding surface, which is often a combination of tracking error and its derivative and/or integral function. Then, the

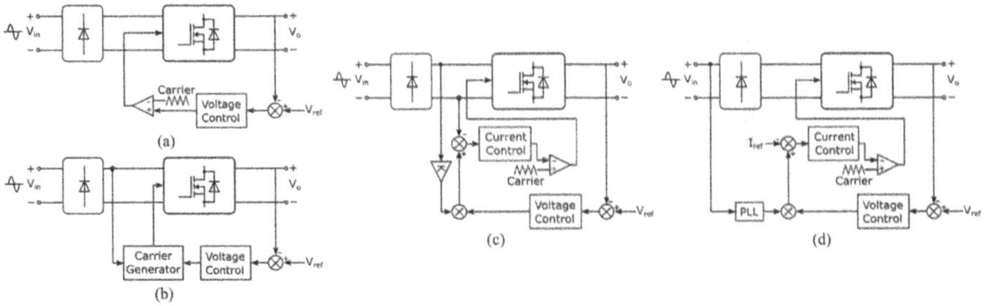

Fig. 4 An overview of typical control types used in PFC converter [18]: (**a**) type I, (**b**) type II, (**c**) type III, and (**d**) type IV

controller's task is to ensure that the system slides to the equilibrium point by using the sign (signum) function of the sliding surface. A key limitation of this approach is that SMC and/or HOSMC are prone to chattering or high-frequency ripple. This may cause a deterioration in the power quality and reduce the component lifetime. In the DEC approach, a tunable decreasing rate is used to ensure the convergence of the state error function [9]. However, the use of derivatives in synthesizing the control law makes this approach sensitive to measurement and/or switching noise. Unlike other nonlinear controllers, conventional FLC does not require any mathematical model and can be tuned in a model-free manner [14, 21]. However, the tuning process is not systematic.

Linear control techniques such as average current control, predictive control [8, 14, 16, 22], and optimization-based parameter tuning of linear control [bee colony optimization (BCO)] [15] are applied in type III control methods. Some of these approaches require precise information about the system parameters. As the converter ages with time, this value will also change, which will deteriorate the control quality. Finally, phase-locked loop (PLL)-based control methods can be classified into type IV. Here, PLL is used to extract information about the grid voltage, and this approach is particularly useful when operating in a distorted grid. Nevertheless, the implementation of software PLL adds an additional computational burden for the real-time implementation in low-cost microcontrollers. A comparative review of control methods for various PFC converters is provided in Table 1.

4 Conclusion and Perspectives

LED drivers are essential to power LEDs for smart farming. Enhancing the efficiency of these drivers is very important to make smart farming energy efficient, reduce carbon emission, and increase the economic viability. In this chapter, a review of topologies and control methods of PFC converters for driving smart farming LEDs is presented. Some common circuit topologies are first discussed.

Table 1 A summary of LED driver topologies and control methods

	Topology		No. of sensors		Controller			Power quality		Complexity	
References	Rectifier	PFC	V	I	Voltage	Current	Modulation	THD	PF	Implementation	Tuning
[7]	DB	Boost	2	2	PI	SMC	PWM	4.85%	0.9990	Medium	High
[8]	DB	Boost	2	0	PI	Predictive	PWM	3%	0.9996	High	Low
[9]	DB	Boost	2	1	PI	DEC	PWM	2.834%	0.9987	Medium	Medium
[10]	DB	I-Boost	2	0	PI	PI	DPWM	1.21%	0.9998	High	High
[11]	DB	I-Boost	2	2	PI	SMC	DPWM	3.43%	0.9993	High	Medium
[12]	DB	Flyback	2	1	–	PI	PWM	NI	≈0.985	High	Low
[13]	Bridge less Ćuk		1	2	PI	PI	PWM	2.43%	0.998	Low	High
[14]	DB	STC	4	1	FLC	Predictive	PWM	3.7%	0.998	High	High
[15]	DB	STC	2	1	PI	BCO	PWM	3.5%	0.990	High	Low
[16]	Totem–Pole Boost		2	2	PI	Predictive	PWM	3.75%	0.997	High	High
[17]	Totem–Pole Boost		2	0	PI	PI	PWM	1.68%	0.9998	Medium	Low
[20]	DB	Boost	2	1	PI	Hysteresis	PWM	2.86%	0.9995	Low	Low
[22]	DB	Boost	2	1	Predictive	HOSMC	PWM	4.4%	0.992	High	High
[21]	DB	Boost	2	1	FLC	Hysteresis	Hysteresis	1.78%	NI	High	Medium
[23]	DB	Boost	2	0	PI	PR	PWM	4.3%	0.996	High	High

NI no information

Additionally, the control methods are critically reviewed, and a table-based comparison summary is presented. Based on the presented summary, it is found that the reported THD value varies very widely between 1.21% and 4.85% depending on the topology and loading condition.

The choice of method used to control the PFC converter is useful in reducing the THD of the system. This is very important as this can directly reduce the power loss and make these converters grid-friendly. The findings from the literature suggest that linear current controllers often result in lower THD compared to the nonlinear controllers. As such, linear control methods can be further improved to reduce the THD value. However, linear controllers often have slow dynamic responses. To address this issue, one potential solution could be using a combination of linear and nonlinear control techniques for the current controller. Conventionally, tracking error is directly passed to the PI controller. However, adding some sort of nonlinear filter before the PI controller can effectively enhance the convergence speed of the tracking error. The other solution could be the use of high-switching frequency for the power switches, which will result in a reduction of the size of passive components.

PLL-based control method is widely used to make the LED drivers resilient to distortions in the grid [23]. Here, the reported THD is significantly higher compared to other techniques. One reason could be the lack of using the current sensor. Moreover, the considered PLLs mostly use frequency-fixed configuration. This can play a role in having a high THD value. To address this issue, an innovative PLL method that is easy to implement and robust in nature needs to be developed. Recently, quasi type I PLL has become popular in the literature as it involves only one gain in the loop filter. Hence, this can be an interesting future research direction, especially the integration of this PLL in PFC converter control. In the sensorless scheme, the current is reconstructed from the voltage signal. However, the current is noisier and has faster dynamics compared to the voltage. As such, more work needs to be considered in this area to improve the estimator performance. Therefore, developing an innovative and noise robust current estimation method can be considered as a potential future work to lower the THD and improve the overall efficiency of the system. The review presented in this chapter will help researchers make informed decisions when choosing or designing LED circuit topologies and control methods in order to improve the overall energy efficiency in smart farming systems.

Acknowledgments This work was supported by the Coventry University Global Challenge Research Fund (GCRF), Fully Funded Doctoral Studentship Scheme to M.L.P., and the Royal Society (RGS/R2/180195) to M.F. H.A. acknowledges the support by Sêr Cymru II 80761-BU-103 project by Welsh European Funding Office (WEFO) under the European Regional Development Fund (ERDF).

References

1. Kozai T 2012 Sustainable plant factory: closed plant production systems with artificial light for high resource use efficiencies and quality produce *Int. Symp. on Soilless Cultivation* 1004 pp 27–40
2. Pereira J, Mouazen A M, Foo M and Ahmed H 2021 A framework of artificial light management for optimal plant development for smart greenhouse application *PLoS One* **16**(12): e0261281
3. Pay M L, Kim D W, Somers D E, Kim J K and Foo M 2022 Modelling of plant circadian clock for characterizing hypocotyl growth under different light quality conditions *in silico Plants* **4** (1):diac001
4. Javed S, Issaoui L, Cho S and Chun H 2021 Utilization of LED grow lights for optical wireless communication-based rf-free smart-farming system *Sensors* **21**(20): 6833
5. Ali M, Orabi M, Ahmed M E and El Aroudi A 2011 Design considerations of a single-stage LED lamp driver with power factor correction *2nd Int. Conf. on Electric Power and Energy Conversion Systems (EPECS)* (Sharjah, United Arab Emirates:IEEE) pp 1–6
6. Yadlapalli R T, Narasipuram R P and Kotapati A 2020 An overview of energy efficient solid state LED driver topologies *Int. J. of Energy* Research **44**(2) 612–30
7. Mohanty P R and Panda A K 2017 Fixed-frequency sliding-mode control scheme based on current control manifold for improved dynamic performance of boost PFC converter *IEEE J. of Emerging and Selected Topics in Power Electronics* **5**(1) 576–86
8. Zaky M S, Azazi H Z and Touti E 2018 PFC control for LED lamp driver using sensorless predictive current controller *Engineering,* Technology *& Applied Science Research* **8**(5) 3373–79.
9. Mohanty P R and Panda A K 2016 A nonlinear control scheme based on dynamic evolution path theory for improved dynamic performance of boost PFC converter working on nonlinear features *ISA Transactions* **65** 254–61.
10. Pahlevani M, Pan S, Eren S, Bakhshai A and Jain P 2014 An adaptive nonlinear current observer for boost PFC AC/DC converters *IEEE Transactions on Industrial Electronics* **61**(12) 6720–9
11. Marcos-Pastor A, Vidal-Idiarte E, Cid-Pastor A and Martinez-Salamero L Interleaved digital power factor correction based on the sliding-mode approach *IEEE Transactions on Power Electronics* **31**(6) 4641–53
12. Zhang J, Zeng H and Jiang T 2012 A primary-side control scheme for high-power-factor LED driver with TRIAC dimming capability *IEEE Transactions on Power Electronics* **27**(11) 4619–29
13. Marimuthu G, Umamaheswari M G, Komathi C and Lekshmi Sree B 2022 Design and implementation of modified type-iii bridgeless Cuk LED driver for smart lighting applications *Electrical Engineering* **104**(4) 2221–36
14. Bouafassa A and Fernandez-Ramirez LM 2022 A DSP-based implementation of fuzzy logic and predictive current control for a Sheppard-Taylor power factor correction converter *Int. J. of Circuit Theory and Applications* **50**(3) 812–26
15. Karaarslan A 2013 The implementation of bee colony optimization algorithm to sheppard-taylor PFC converter *IEEE Transactions on Industrial Electronics* **60**(9) 3711–19
16. Chen J, Tai W, Xun B, Gong C and Chen J 2022 Improved control strategies for totem-pole PFC with true full range ZVS operation *IEEE Transactions on Industrial Electronics* **70**(3) 2419–30
17. Chandwani A, Dey S and Mallik A 2021 Parameter variation tolerant robust current sensorless control of a single-phase boost PFC *IEEE J. of* Emerging *and Selected Topics in Industrial Electronics* **3**(4) 933–45
18. Lamo P, Castro A, Sanchez A, Ruiz G A, Azcondo F J and Pigazo A 2021 Hardware-in-the-loop and digital control techniques applied to single-phase PFC converters *Electronics* **10**(13) 1563

19. Lopez F, Lopez-Martin V M, Azcondo F J, Corradini L and Pigazo A 2019 Current-sensorless power factor correction with predictive controllers *IEEE J. of Emerging and Selected Topics in Power Electronics* **7**(2) 891–900
20. Abd El-Moniem M S, Azazi H Z and Mahmoud S A A current sensorless power factor correction control for LED lamp driver *Alexandria Engineering J.* **53**(1) 69–79.
21. Sawachan H, Chaicharoenaudomrung K and Pakdeeto J 2021 Power factor correction of a single phase AC/DC boost converter using fuzzy logic controller *18th Int. Conf. on Electrical Engineering/Electronics, Computer, Telecommunications and Information Technology (ECTI-CON)* (Chiang Mai, Thailand: IEEE) pp 671–4
22. Bouafassa A, Rahmani L and Mekhilef S 2015 Design and real time implementation of single phase boost power factor correction converter *ISA Transactions* **55** 267–74
23. Lamo P, Lopez F, Pigazo A and Azcondo F J 2018 Stability and performance assessment of single-phase t/4 PLLs with secondary control path in current sensorless bridgeless PFCs *IEEE J. of Emerging and Selected Topics in Power Electronics* **6**(2) 674–85

Part II
Wind Power Generation and Ocean Energy

Monitoring Wind Turbine Blade Using Interferometric Radar

Gokhan Kilic

1 Introduction

A wind turbine blade is made up of two faces (on the suction side and the pressure side), which are attached and made stiffer by either one or more integrated (shear) webs connecting the upper and lower parts of the blade shell or by a box beam (box spar with shell fairings) (Fig. 1). [1, 2].

Various reports explaining the wind turbine's condition pinpointed several contributing factors. These were most notably the dynamic effects of weather conditions, cracks in the wind turbine, rebar corrosion (exacerbated by weather conditions), and the absence of comprehensive maintenance work [1–4]. Although certain nondestructive testing (NDT) processes call for very minor apparent structural degradation, its testing is frequently employed to evaluate the condition of structures [1, 5]. Sensors can detect the whole spectrum of vibrational frequencies expected to be present during wind turbine blade maneuvers [1, 2, 4]. NDT is frequently used by wind turbine engineers because of its ability to identify structural flaws and weaknesses as well as the structure's true configuration. Recent advancements in NDT have made it possible to precisely remotely inspect the condition of wind turbine blades using instruments that can capture real vibrations. By eliminating the requirement for wind turbine engineers, this sort of remote and autonomous assessment and verification improves data collecting and safety [2–7].

One type of NDT technique is the IBIS-S radar system, which transmits and receives electromagnetic waves exploiting electromagnetic interference between coherent waves. This system gives researchers data accurately to one-hundredth of a millimeter for both dynamic and static movement and vibration of structures, such as towers, dams, and bridges. When vertical bending motions outweigh torsional

G. Kilic (✉)
Izmir University of Economics, Izmir, Turkey
e-mail: gokhan.kilic@iu.edu.tr

Z. Peng (ed.), *Trends and Technological Challenges in Green Energy*, Green Energy and Technology, https://doi.org/10.1007/978-3-031-55579-4_5

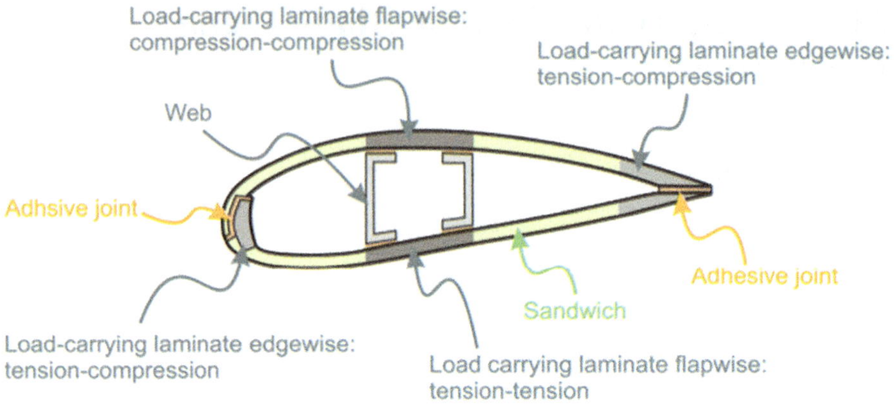

Fig. 1 Schema of the section of the blade [1]

movements, IBIS-S can identify structural displacement [1, 6]. This technology is essential for strengthening our comprehension of wind turbine dynamics, especially the practical challenges associated with monitoring wind turbine dynamics while concurrently detecting wind forces [7].

The current study suggests a method to prevent these deficiencies using self-organized ONNs (self-ONNs) with generative neurons that can optimize each link's nodal operator during the training phase to increase network diversity while also improving computing efficiency.

2 The IBIS-S Principle and Monitoring

IBIS-S monitoring is an advanced and precise NDT technology for detecting structural movement and vibration with millimeter accuracy. As long as vertical bending motions outweigh torsional ones, an IBIS-S system is able to monitor a structure's displacement. Additionally, the interferometric sensor module can send, receive, and analyze electromagnetic waves for the source of electromagnetic reflection waves. Despite these advantages, there is a scarcity of studies analyzing dynamic behavior in wind turbines, the continuous monitoring of which is challenging [7].

The IBIS-S system can provide data such as wind turbine blade displacement information via the geometric dispersion of movement and dynamic indicators. The IBIS-S system performs very well since it continuously generates electromagnetic waves with a bandwidth rate of up to 100 Hz [8]. Interferometric analysis offers information about the object's movement accurately to the closest 0.01 mm by correlating the level information gathered at various time intervals of the waves received from the item [8]. Equation (1) predicts the movement in the form of wave

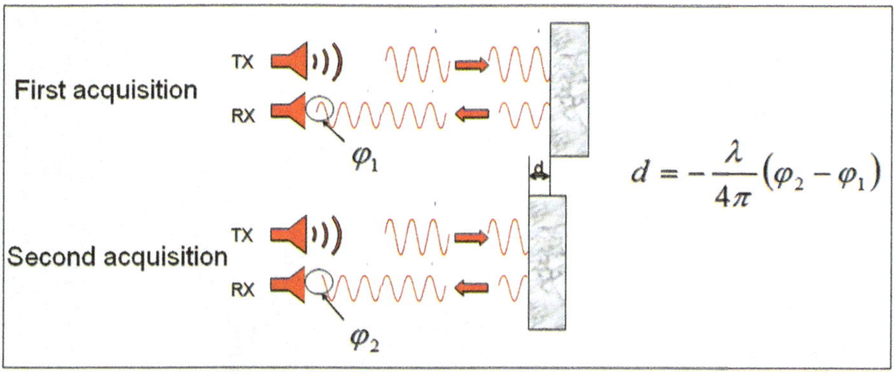

Fig. 2 Source of interferometric quantity [8]

transmission [8]. Trigonometric calculations and radial movement data can be used to determine movement in an appropriate coordinate system (Fig. 2).

$$dp = \frac{\lambda}{4.\pi}.\Delta\varphi \tag{1}$$

where

dp = measured displacement
λ = wavelength
$\Delta\varphi$ = carrier wave phase shift

Variance interferometry analyzes the phase evidence of backscattered electromagnetic waves collected at various intervals [8] and is used to determine the displacement responses at several places in various acquisition periods. To determine the phase transference observed by the radar units during various acquisition times, IBIS-S broadcasts and receives electromagnetic waves. Equation (2) provides examples for electromagnetic signal wavelength (λ), phase shift ($\Delta\varphi$), and circular movement (dp) [8, 9].

$$dp = \alpha \ \frac{\lambda}{4\pi}\Delta\varphi \tag{2}$$

When a target shifts, it commonly returns a point shift in the radar signals at different acquisition times. Equation (2) establishes a connection between radial movement d_r and phase shift $\Delta\varphi$, where λ is the wavelength of the electromagnetic signal [7]. Radial movement is the movement along the wave's line of transmission.

The IBIS-S system consists of a radar device with two spinning horn antennas mounted on a pedestal. The radar is controlled with IBIS-S software, which runs on a standard laptop and is powered by a 12 V, 12 Ah battery (Fig. 3).

Fig. 3 View of IBIS-S system in use

IBIS-S is not transportable even though it is simple and quick to apply in all weather conditions, and enables almost continual monitoring of the whole structure's static and dynamic displacements. IBIS-S can detect static or dynamic motions, even in the presence of obstructions to the electromagnetic waves such as water, trees, fences, cables, or other structures (Fig. 4).

The primary flaw is IBIS-S originating in the area between the radar and the reflectors being covered. It is imperative that no additional reference points are allowed to enter this area while the survey is being conducted. Even though it cannot directly identify structural flaws, IBIS-S analyzes wind turbine blade behavior under stress, which is an inherent wind turbine blade vibration problem [8].

3 Results

IBIS-S is capable of detecting line-of-sight deformations as small as 0.01 mm and vibrations up to 100 Hz. It generates a one-dimensional image with a range determination of up to 0.5 m and has a maximum range of approximately 500 m. Interferometric radar testing can produce dynamic images, with a sampling rate high enough to track the movements of objects such as wind turbine [7].

Data were gathered from a wind turbine using the synthetic aperture radar (SAR) method, which enables radars with high physical antenna ray widths to accomplish excellent azimuth angle determination. It was able to build an artificial antenna with a beam width opposite from the relative movement between the backdrop and the

Fig. 4. Monitoring wind turbine blade via IBIS-S system

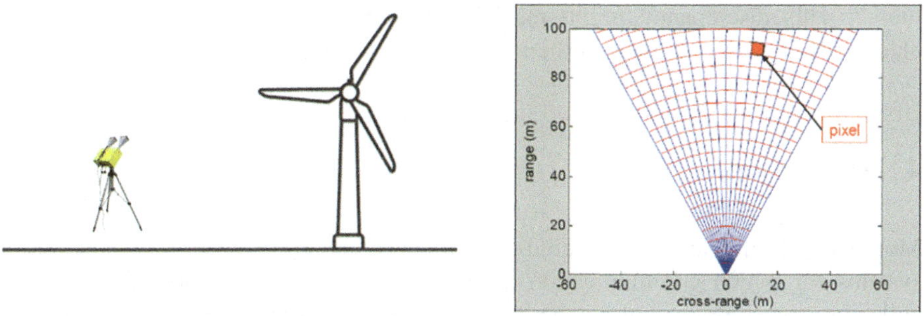

(a) (b)

Fig. 5 (**a**) Dynamic monitoring of the wind turbine; (**b**) the spatial resolution grid

radar since the acquisitions are logical [8]. The combination of SF-CW and SAR techniques allowed the radar image to be separated into pixels with diameters of 0.5 m in range and 4.5 (10-3) m in cross-range, where r is the distance between the radar and object (Fig. 5).

When comparing precise observed points, it is essential to consider the item's shape. If the range and cross-range information in the horizontal level are appropriately sorted and the item has a vertical surface, the amount of scatters is minimal, but on a sloped surface, it rises significantly [9]. After training five CNNs, the vibration detection system was evaluated using eight reflector instances (Fig. 6).

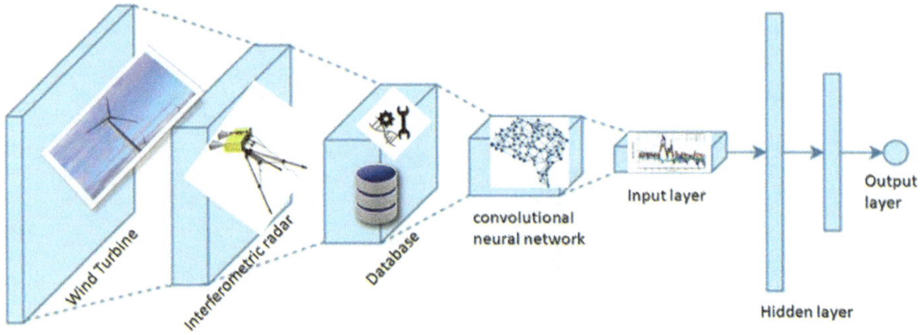

Fig. 6 CNN architecture

The corner reflector 3, 4, 5 and 6 deflection is depicted in Fig. 7, along with the displacement with frequency that was produced by spectral analysis of the raw signals using the CNN-based approach.

The results unmistakably show targeted displacement occurring, often in different wind turbine, proven by reflectors' in-the-moment observation of the wind turbine's behavior, despite a tiny discrepancy between the IBIS-S data and that depicted in Fig. 7. This discrepancy might result from either differences in the horizontal placement of each system or in the measurement techniques.

4 Discussion

More study in this area will undoubtedly increase the effectiveness of CNN-based systems for which communication is essential. In this investigation, the critical collecting transmission link to the base unit consisted of eight points. The advantages of multiple nodes communicating with the base unit might be investigated in future research [1, 2, 5, 8, 9]. The limitations imposed by the structure's design are one drawback of the evaluation method for wind turbine blades described here. There are fewer places where a weaker signal may be created, and these should be avoided. Examples of such features are solders, bolts, and connections [10–12].

Eight reflector data in total were utilized for training, with reflector 7 being used for testing. The test findings showed that the new CNN had a 95% accuracy rate. The criteria for correct value differences of less than 0.2 cm between the absolute value of the actual vibration is used. While BPNN was effectively employed to gather geographical data, CNN remains the more effective approach. Both different types of vibrations in the data and distinct observations within each reflection test may be identified using CNN. CNN uses a single filter on all of the data to produce a feature map (Fig. 6), meaning that there is no need for the crucial part of BPNN's manual

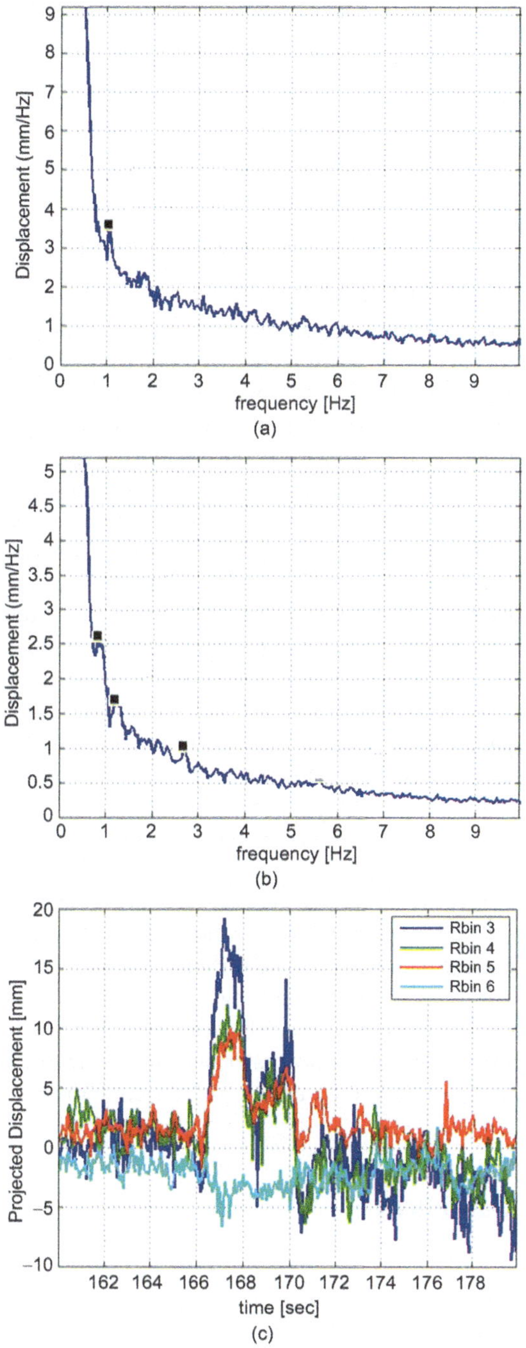

Fig. 7 (a) Reflector 4: displacement with frequency; (**b**) reflector 5: displacement with frequency; (**c**) reflector 3, 4, 5 and 6: displacement with time

feature extraction. As shown, artificial intelligence is one method for evaluating measures in the early phases of structural health, among others.

5 Conclusions

IBIS-S was used in this work to present an innovative real-time health diagnosis approach. To address the deformation safety hazards caused by wind turbine blade vibration, CNN was used to provide a detailed view of the main data processing flow. At eight reflector monitoring distances, simultaneous reliable time-series deformation measurements were successfully acquired. Thus, IBIS-S may be used to track the dynamics of wind turbine blade monitoring, enabling currently unrivaled measurements in three dimensions and across time. This study demonstrated the accuracy of millimeter-level monitoring provided by the IBIS-S system as well as the efficacy and dependability of this monitoring strategy. The four benefits of the method are as follows: its ability to monitor the wind turbine blade's welfare and condition, the delivery of data support and warnings, the prevention of deformation catastrophes due to wind turbine blade vibration, and the reduction of fatalities and structural damage caused by distortion catastrophes.

In contrast to conventional monitoring techniques, the IBIS-S methodology is a contactless measurement method for a concurrently dynamic, high-precision automated distortion surveying of wind turbine blade vibration. It creates continual space-time distortion data from the region under study, which it uses to accurately calculate the safety hazards.

The outputs of the CNN were the actual measurements taken during the training phase, and these data were entered into the training phase including data from eight reflectors, and the accuracy of the CNN's estimate was evaluated using data from a different reflector. The testing revealed a 95% accuracy rate.

The study presented a unique approach, based on the comparison of dynamic reaction data from a wind turbine structure with a specially designed CNN. The data is processed by this cutting-edge neural network model in a matter of seconds, practically in real time. The outcomes demonstrate the ability of a CNN to estimate correct measurements faster and more effectively. The CNN's prediction accuracy can be improved by adding further components.

References

1. Mishnaevsky, L.; Branner, K.; Petersen, H.N.; Beauson, J.; McGugan, M.; Sørensen, B.F. Materials for Wind Turbine Blades: An Overview. Materials 2017, 10, 1285. https://doi.org/10.3390/ma10111285
2. Zervos, A. Renewable Energy Technology Roadmap 20% by 2020; European Renewable Energy Council: Brussels, Belgium, 2009.

3. Mishnaevsky, L., Jr.; Brøndsted, P.; Nijssen, R.P.L.; Lekou, D.J.; Philippidis, T.P. Materials of large wind turbine blades: Recent results in testing and modelling. Wind Energy 2012, 15, 83–97
4. Watson, J.C.; Serrano, J.C. Composite Materials for Wind Blades. Wind Syst. Mag. 2010, 46–51. Available online: http://www.ppgfiberglass.com/getmedia/0b0f3799-494a-40e8-b67d-b64fc7ee8adc/ WindSystems_PPG.pdf.aspx
5. Garolera, A.C.; Madsen, S.F.; Nissim, M.; Myers, J.D.; Holboell, J. Lightning Damage to Wind Turbine Blades from Wind Farms in the US. IEEE Trans. Power Deliv. 2016, 31, 1043–1049
6. Gokhan Kilic, Mehmet S. Unluturk, Testing of wind turbine towers using wireless sensor network and accelerometer, Renewable Energy, Volume 75, 2015, Pages 318–325, ISSN 0960-1481, https://doi.org/10.1016/j.renene.2014.10.010.
7. Yasuda, Y.; Yokoyama, S.; Minowa, M.; Satoh, T. Classification of Lightning Damage to Wind Turbine Blades. IEEE J. Trans. 2012, 7, 559–566.
8. Cattin, R. Icing of Wind Turbines: Vindforsk Projects, a Survey of the Development and Research Needs. ELFORSK Report 12:13. 2012. Available online: http://winterwind.se/2012/wp-content/uploads/2012/08/Icing_of_windturbines_elforsk_report_12_13_draftversion_120127.pdf
9. Battisti, L. Relevance of Icing for Wind Turbines. In Wind Turbines in Cold Climates, Green Energy and Technology; Springer: Cham, Switzerland, 2015.
10. Gokhan Kilic, Condition assessment of different historic bridges using Non Destructive Techniques (NDT) with FTIR analysis in Izmir after the Samos Island earthquake. Herit Sci. 2022, 10, 111. https://doi.org/10.1186/s40494-022-00746-x.
11. Gokhan Kilic, Wavelet analysis and NDT for condition assessment of historic Masonry Bridge, Structures, Volume 45, 2022, Pages 275–283, ISSN 2352-0124, https://doi.org/10.1016/j.istruc.2022.09.008.
12. Mengal, A.N.; Karuppanan, S.; Wahab, A.A. Basalt Carbon Hybrid Composite for Wind Turbine Rotor Blades: A Short Review. Adv. Mater. Res. 2014, 970, 67–73.

Generation of Fluid Field Around Tidal/Ocean Power Generator with Unidirectional Flow Exploiting Variable Auto Encoder

Ryo Fujiwara, Hiroki Nakagawa, Tsubasa Ebiko, and Makoto Miyatake

1 Introduction

Renewable energy is an essential factor in guaranteeing the sustainability of society. At present, almost all of the energy in the world relies on fossil fuels. At present, nonrenewable energy sources provide 80% of the world's energy [1]. The uses of fossil fuels are caustic to human health and a major contributor to climate change. The search for alternatives is imperative. Renewable energy is expected as a replacement for nonrenewable energy sources.

Machine learning is focused as one of the methods to manage renewable energy. There have been studies of machine learning for forecasting ocean energy, like wave energy and tidal energy. Bento et al. optimized deep neural networks to forecast the wave energy flux and other wave parameters [2]. Aly forecasts tidal current constitution by a hybrid model of wavelet neural network, artificial neural network (ANN), Fourier series based on least squares, and recurrent Kalman filter [3]. Jörges et al. developed a machine learning model based on long short-term memory (LSTM) neural networks for reconstruction and short- and long-term prediction of nearshore significant wave height (SWH) [4]. Meng et al. propose a bidirectional gated recurrent units (BiGRU) network for the prediction of wave height during tropical cyclones (TC) [5]. Machine learning is also used to control ocean power plants. Amundarain et al. controlled a wave plant by ANN and improved performance during voltage dips in the grid [6]. M'zhoughi et al. constructed an ANN-based airflow control system of oscillating water column (OWC) to prevent stalling phenomena [7]. M'zhoughi et al. constructed a rotational speed control system of the turbine in OWC by ANN-based MPPT [8]. Ghefiri et al. forecast the tidal stream generator's rotational speed and blade pitch angle from the flow velocity by

R. Fujiwara (✉) · H. Nakagawa · T. Ebiko · M. Miyatake
National Institute of Technology, Hakodate College, Hakodate, Hokkaido, Japan
e-mail: r_fujiwara@hakodate-ct.ac.jp

Z. Peng (ed.), *Trends and Technological Challenges in Green Energy*, Green Energy and Technology, https://doi.org/10.1007/978-3-031-55579-4_6

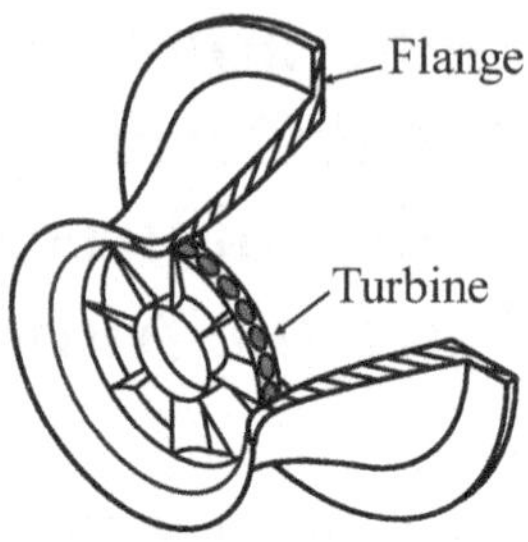

Fig. 1 Schematic illustration of tidal/ocean power generator with "flaring flanged diffuser" (FFD) [28]

multilayer ANN [9]. In addition, machine learning is used to forecast wind velocity and energy [10–13]. In photovoltaic (PV) systems, machine learning is also used as a battery smoothing controller using total sky images [14]. The load, where to use energy, is also forecasted by machine learning [15–24].

Our study focuses on ocean energy, especially the energy from ocean/tidal flow. According to the review article on ocean power technology [25], the potential estimate for tidal energy was 3 TW. Ocean energy plants have merits: the equipment space does not require land use, which can be expensive or scarce, and the energy density is higher than the air. In Japan, there are several developments to harness energy from the ocean [26, 27].

The Tsugaru Strait, in the northern region of Japan, is an area that has attracted attention for the utilization of tidal/ocean energy. Unidirectional tidal and ocean flow with high-energy density drifts between the main island and Hokkaido Island. Energy harvesting, equipping multiple tidal/ocean power generators in the Tsugaru Strait, has been shown to be effective. We propose a tidal/ocean power generator utilizing a flaring flanged diffuser (FFD), as shown in Fig. 1, to harness the power effectively. The flaring shape and the flange increase the tidal/ocean flow velocity, and stable energy can be extracted. We aim to construct a machine learning architecture, as shown in Fig. 2. We have forecasted the design values from the fluid field around the tidal/ocean power generator utilizing FFD by a deep neural network [28].

However, the prediction/forecast of some data can only contribute to understanding a piece of natural phenomenon. Rich data generation is needed to predict/forecast situations around the renewable power generator that interacts with natural phenomena. To obtain rich data, generative deep learning is focused. The data generation aims to identify the probabilistic model for the production of data.

In this chapter, we aim to generate rich data useful for understanding the essence of natural phenomenon. This chapter aims to generate a fluid field around a tidal/ocean power generator. Variable auto encoder (VAE) is used to reconstruct, encode and decode, the fluid field data. The original data of the generation is obtained from the measurement data in the experiment. The architecture of the generative deep neural network is constructed by exploiting the decoder in the learned VAE. The reconstruction of the fluid field is performed, and the performance is evaluated.

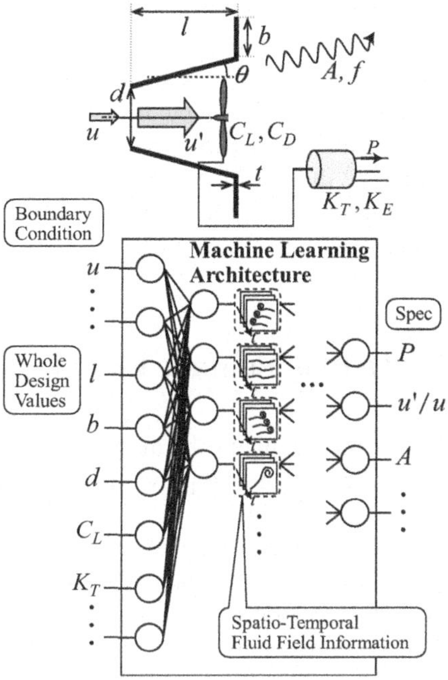

Fig. 2 The concept illustration of machine learning architecture to optimize designing of tidal/ ocean power generators [28]

2 Mechanism of Variable Auto Encoder

Variable auto encoder (VAE) is a generative deep learning model first introduced by Kingma et al. [29]. The following year, Rezende et al. expanded on the model by increasing its effectiveness [30]. VAE is a deep learning network that encodes high-dimensional input data to latent space, decodes to high-dimensional data, and reconstructs the data. Unlike auto encoder (AE), VAE determines the distribution on the latent space with regularization of the shape of the distribution.

VAE consists of an encoder and a decoder. In this section, the image generation process is explained as an example. The encoder transforms the input image data into latent space by neural networks such as convolutional neural networks. Generally, image data is an array of dimensions that are horizontal axis, vertical axis, and color (such as RGB). The latent space is an array of dimensions with much fewer dimensions than the input image data. The encoder is learned to make each image data correspond to each point in the latent space. In the learning, the distribution of the points in the latent space is regularized to match standard normal distribution. The encoder outputs not only the coordinates in the latent space but also the variance. The coordinates and the variances plot the points in the latent space. The decoder transforms the point in the latent space into image data by neural networks such as transposed convolutional neural networks, reconstructing (generating) the image data.

Reconstruction loss is calculated between the original and generated image data. Furthermore, the weights of the encoder and the decoder are optimized to minimize the reconstruction loss. At the same time, Kullback–Leibler divergence (KL divergence) is calculated to make the distribution match the standard normal distribution. And the encoder and the decoder learn (optimize their weight) to minimize the KL divergence. From the above operations, the VAE can obtain the distribution of the points in the latent space corresponding to the original image and appropriately reconstruct the image.

3 Method to Make Fluid Field Data

This section explains the method to make fluid field data (original data). The details are explained in our previous report [28]. In this chapter, the outline is briefly explained.

First, the diffuser model is manufactured as follows. Figure 3 shows the shapes of the FFD model. In this study, we adopted a flaring flanged diffuser (FFD) as the fluid velocity accelerator. The concept of FFD was first proposed by Mehmood et al. [31].

Two-dimensional FFD models are manufactured by changing flange length h, outlet diameter D', and full-length L based on the basic shape of the diffuser, No. 0, as shown in Fig. 3. These models are made from a clear acrylic resin. The

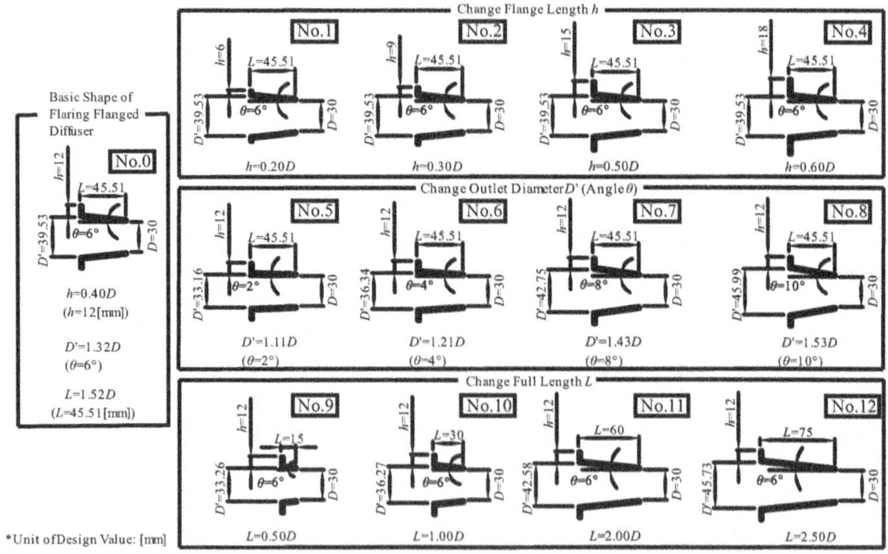

Fig. 3 Shapes of flaring flanged diffuser (FFD) model. No. 0 is the basic shape of the diffuser. Flange length h, outlet diameter D' (that depends on the opening angle), and full-length L are changed, respectively [28]

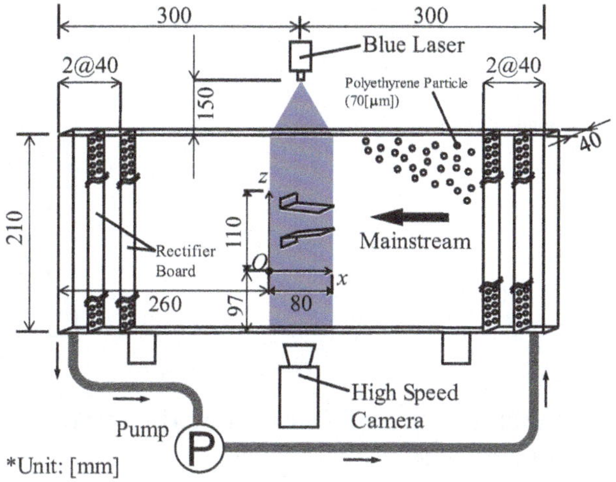

Fig. 4 Experimental setting with steady flow circulation tank [28]

width of the model is 40 mm, which is the same as the width of the tank used in experiments. Then, the model experiments are conducted in a steady flow circulation tank, as shown in Fig. 4. The circulation tank is made of acrylic resin and has a rectifier board with holes in its inlet and outlet.

Fluid field data is measured by PIV as follows. First, mainstream velocity is modified by generating steady flow circulation without a diffuser. The velocity is modified based on the flow conditions of the Tsugaru Strait.

Polystyrene particles with a diameter of 70 μm (Thermo Fisher, 4230A) are dispersed in the tank to conduct PIV. A blue laser irradiates the particles with 1 W power to observe their motion. A high-speed camera with a frame rate of 200 fps (Direct, HAS-L1) is used to record the movies of the motion of the particles irradiated by a blue laser. The fluid field data is obtained by analyzing the movies (continuous images) by PIV. ImageJ, the image processing software, is used in PIV analysis. Then, the FFD model is installed, as shown in Fig. 4. The fluid field data is obtained by observing flow conditions around the model using PIV. The coordinate is set as Fig. 4. The origin O is at the bottom of the downstream, the positive direction of the x-axis points to the upstream, and the positive direction of the z-axis points upward in a vertical direction.

A three-dimensional array with the shape of $2 \times 48 \times 48$ is adopted to store the fluid field data. The first element indicates the direction of velocity, u, and v. The second and third elements indicate the x and z positions. The quantity of movie frames and shape variations is also considered. The shortest movie has 4900 frames. Thirteen kinds of diffuser models were made. Thus, the whole tensor shape is $2 \times 13 \times 4900 \times 48 \times 48$. The details of the fluid field tensor shape are shown in Fig. 5. In this chapter, we used only horizontal velocity u. Thus, the tensor shape that we used is $1 \times 13 \times 4900 \times 48 \times 48$.

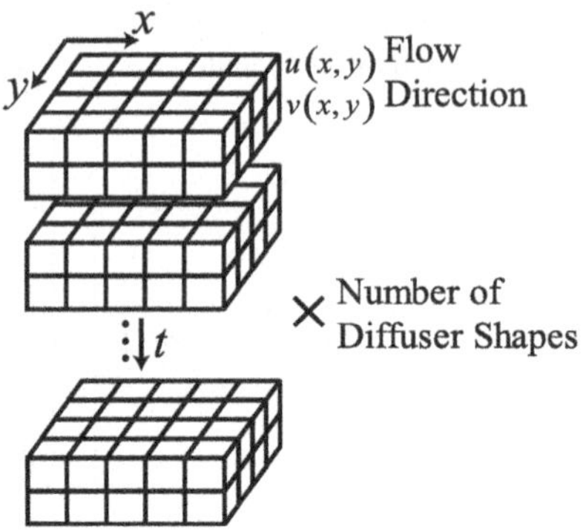

Fig. 5 The details of the fluid field tensor shape. The tensor has five dimensions. In deep learning, only horizontal velocity u is used [28]

4 Architecture of Deep Neural Network to Generate Fluid Field

This section explains the architecture of the deep neural network to generate a fluid field. There are two steps in implementing the deep neural network. First, VAE, the encoder and the decoder, learns to reconstruct the fluid field. Then, the relation between the design values of FFDs and the points in the latent space is learned by fully connected neural networks, and a point of latent space is decoded into a fluid field by the decoder. This architecture, which transforms the design value data into a fluid field via the latent space, can generate the fluid field.

4.1 The Architecture of VAE

The VAE architecture that encodes and decodes fluid field is shown in Fig. 6.

The original image is the spatial distribution of horizontal velocity in a fluid field. This fluid field image is stored in an array of 48 × 48. The latent space is a two-dimensional space. From the original image's shape and the latent space's dimension, the encoder's hyperparameters are determined as shown in Fig. 6. The quantity of the original fluid field image is 50,000. The order of the image is randomly lined up. The velocity value is normalized to let its minimum and maximum values be 0 and 1, respectively. This normalization is conducted to input the transposed convolutional neural network described below. The dimension of the latent space is two since there was no specific difference in the output images even if the dimension is several dozens. Then, the decoder transforms the points in the latent space into the output (generate) image by transposed convolutional neural networks. Since the size of the generated image and that of the original image is the same, the decoder's hyperparameters are determined as in Fig. 6.

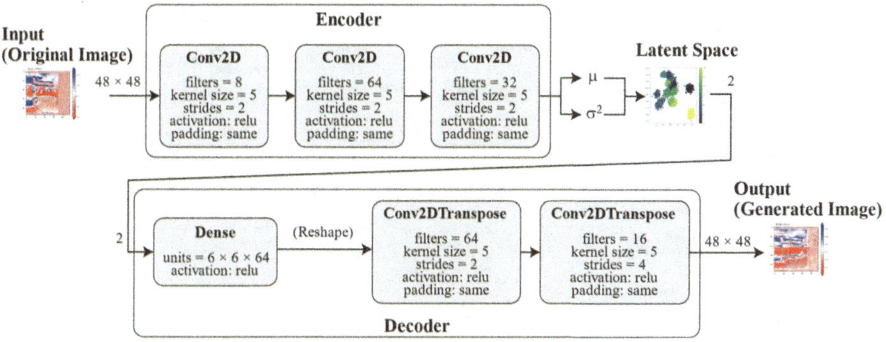

Fig. 6 The VAE architecture that encodes and decodes fluid field. The fluid field image is stored in a 48 × 48 array. The latent space is a two-dimensional space. From the original image's shape and the latent space's dimension, the VAE's hyperparameters are determined

The loss functions are used to optimize the weight of the encoder and the decoder, and reconstruction loss and regularization loss are calculated as follows. In the reconstruction loss, binary cross entropy $B(T, z)$ is calculated as

$$B(T, z) = -[T\log(z) + (1 - T)\log(1 - z)], \tag{1}$$

where T is the true value (the value in the original data), and z is the output value (generated value from the decoder). In the regularization loss, KL divergence $D_{\text{KL}}(\mu,\sigma)$ is calculated as

$$D_{\text{KL}}(\mu, \sigma) = -\frac{1}{2}\left[1 + \log\left(\sigma^2\right) - \mu^2 - \sigma^2\right], \tag{2}$$

where μ and σ denote the mean value and variance value output from the encoder. The total loss of the VAE is calculated as $B(T, z) + D_{\text{KL}}(\mu,\sigma)$. The optimization (minimization of the loss function) method is Adam [32]. The learning batch size is 512. The learning epoch is 30. The VAE can encode and decode the fluid field from the above conditions and learning methods.

4.2 *The Architecture of Generative Deep Neural Network Exploiting VAE Decoder*

By exploiting the VAE decoder, the fluid field can be generated from the design values of FFD. The architecture of the generative neural network is shown in Fig. 7.

There are two steps to transform design values to fluid field. First, a design value is transformed into a point in the latent space. Then, the point in the latent space is transformed into fluid field by the decoder in VAE. The transformation from the

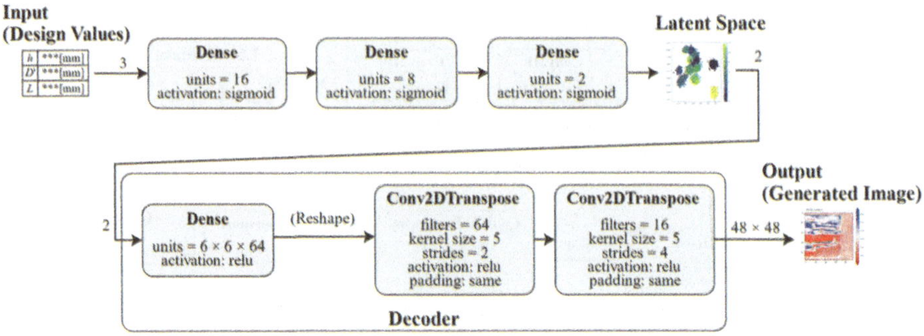

Fig. 7 Architecture of the generative deep neural network for the fluid field. There are two steps to transform design values to fluid field. First, a design value is transformed into a point in the latent space. Then, the point in the latent space is transformed into fluid field by the decoder in VAE

design value into the point in the latent space is conducted as follows. The diffuser's shape is determined by three parameters: flange length h, outlet diameter D', and full-length L. The inlet length D is treated as a constant. The set of these three parameters (h, D', L) is called the design value. The design value, an array with three values, corresponds to the point in the latent space, an array with two values. A fully connected deep neural network is used to correspond these data. The hyperparameter of the network is shown in Fig. 7. Mean squared error is used as the loss function. Adam is used as the optimizer. The learning batch size is 512. The learning epoch is 20. The decoder in the VAE conducts the transformation from the point in the latent space into fluid field. The learning of the decoder has been completed by following the operations in Sect. 4.1.

The fully connected neural network and the decoder are obtained from the above conditions and learning methods. The architecture of the generative neural network is obtained by connecting these two networks.

5 Results and Discussions of the Generation of Fluid Field Exploiting VAE

First, the VAE is learned to reconstruct fluid field. The distribution in the latent space is shown in Fig. 8. The color of the points indicates the number of design value. Horizontal and vertical axes indicate the coordination in the latent space. This distribution is determined depending on the initial state of the weight of the networks. Thus, this distribution changes almost arbitrarily. The points of the latent space can distinguish the difference in the number of design values. However, the points of design value No. 0 and No. 12 are located at distances from other points. These distances occur because of the difference in the diffuser's shape and the subtle difference of conditions in the measuring.

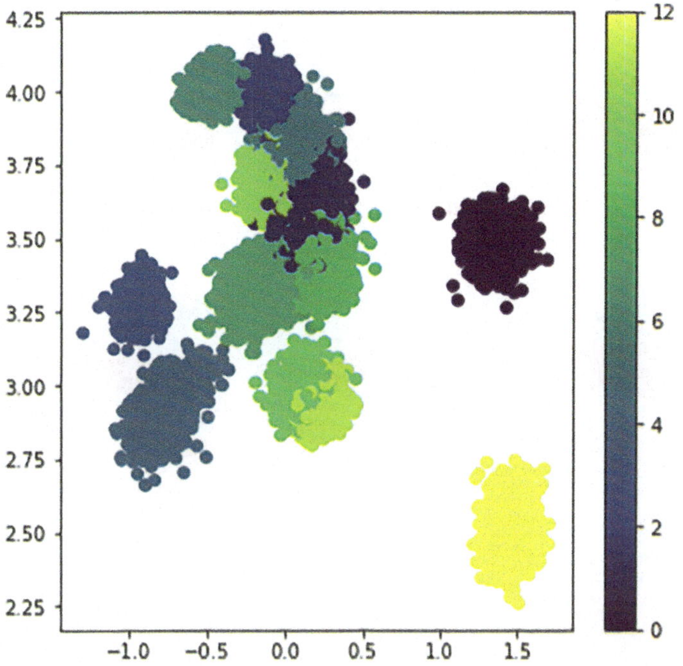

Fig. 8 The distribution of the points in the latent space. The color of the points indicates the number of design value. Horizontal and vertical axes indicate the coordination in the latent space. This distribution is determined depending on the initial state of the weight of the networks. Thus, this distribution changes almost arbitrarily

Then, the encoder in the VAE is evaluated. The nine points are extracted using the minimum, middle, and maximum values in the coordinate of the points. The nine fluid fields are generated from these nine points. Figure 9 shows the generated fluid fields. The red region indicates the high-velocity region, and the blue one indicates low-velocity region. From Fig. 9, it is confirmed that the decoder can generate a fluid field close to the original fluid field.

Finally, a fluid field is generated from the design value by the architecture of the generative deep neural network. Figure 10 shows the original fluid fields and the generated fluid fields corresponding to the number of design values. The velocity of the original fluid field is averaged over time. In each number of the design value, the fluid field can be almost generated. Although there are subtle differences in the fluid field, the detail, such as the shape of the diffuser and the distribution of fluid field, cannot be recreated. This regularization generally occurs in the generation by VAE. It would occur because the probability distribution to generate an image is fixed as the standard normal distribution, which has only one mode value.

From the above, it is confirmed that fluid field can be generated from the design value. The recreation of the detail in fluid field remains for future work. Two

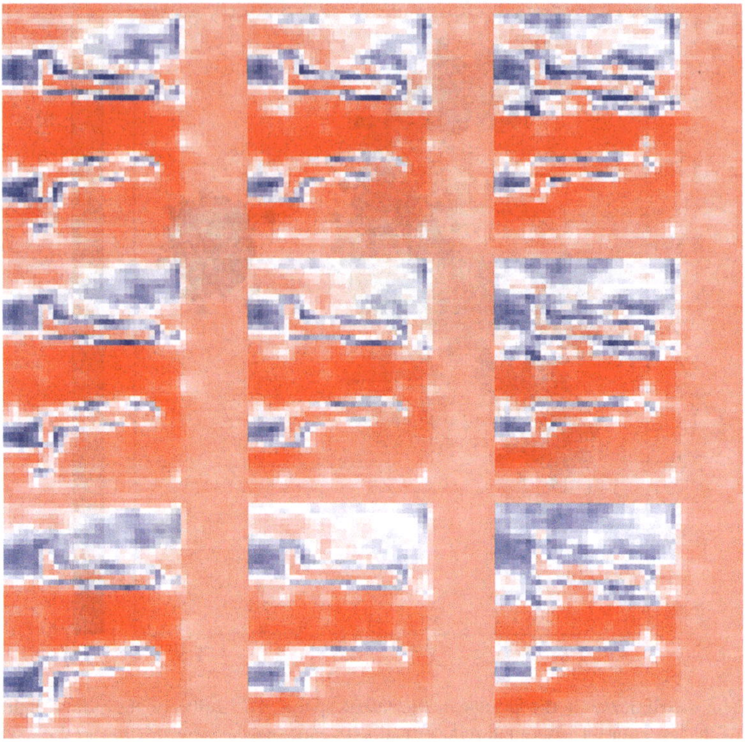

Fig. 9 The fluid fields generated from the points of the latent space. The nine points are extracted using the minimum, middle, and maximum values in the coordinate of the points

methods are expected to solve this problem, increasing the dimension and using GAN. If the dimension of the latent space is increased, the variation of the expression would be richer. Generative adversarial networks (GAN) [33] are also expected to generate fluid field with more rich expression.

6 Conclusion

In order to comprehend the essence of natural phenomena, we produced rich data through this work. Generative deep learning is focused to obtain rich data around the tidal/ocean power generator utilizing flaring flanged diffuser (FFD).

Variable auto encoder (VAE) is used to reconstruct, encode and decode, the fluid field data. The original data of the generation is obtained from the measurement data in the experiment. The architecture of the generative deep neural network is constructed by exploiting the decoder in the learned VAE. The reconstruction of the fluid field is performed, and the performance is evaluated. It is confirmed that the

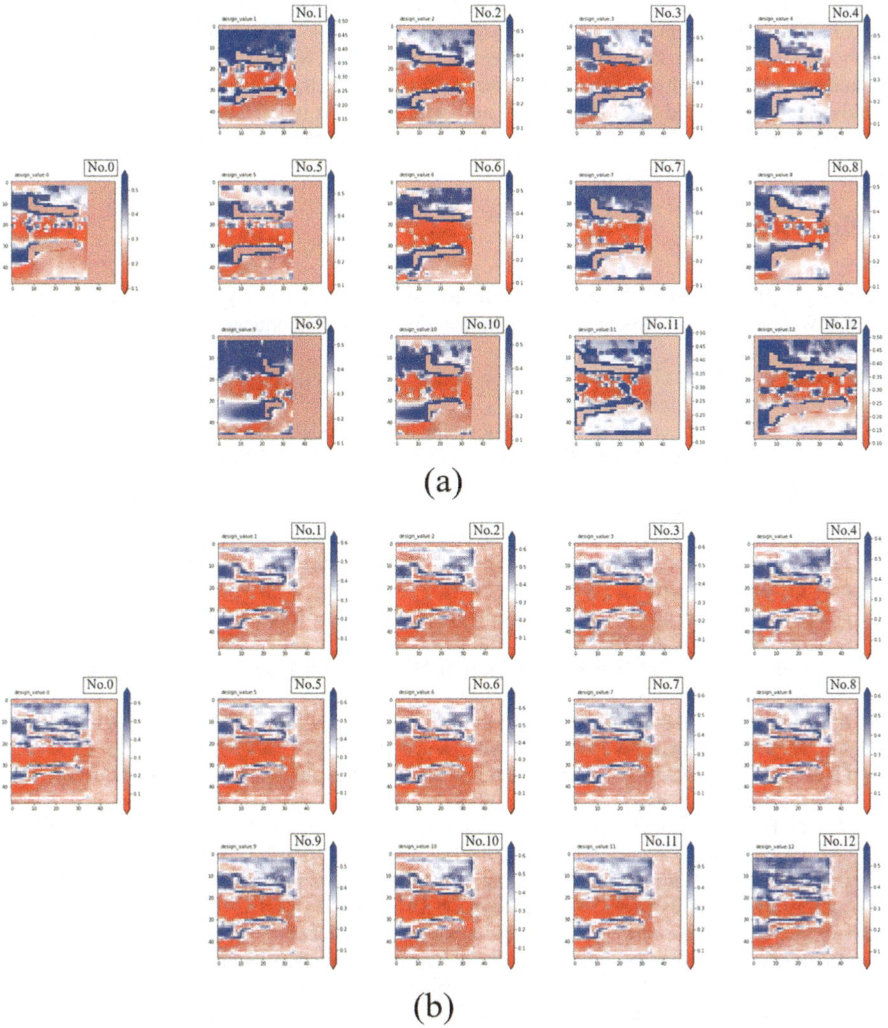

Fig. 10 (**a**) The original fluid fields and (**b**) the generated fluid fields corresponding to the number of design values. The velocity of the original fluid field is averaged over time

generative deep neural network can generate the fluid field close to the original fluid field. The recreation of the detail in fluid field will be addressed in further work. It is expected that there are two methods to solve this problem, increasing the dimension of latent space and using generative adversarial network (GAN).

This proposed method will contribute to understanding the essence of natural phenomena and the development of renewable energy systems. The proposed method can be applied to various natural phenomena, wind, solar, and biomass.

Acknowledgments This work was supported by the JSPS KAKENHI Grant No. 21K14081 and Leave a Nest KOSEN grant Focus Systems award. The authors would like to thank Mr. Robert Anthony Olexa of the Department of General Humanities, National Institute of Technology, Hakodate College, for his helpful advice on English writing. R. F. would like to thank his own family, and Y. K. F. and Y. F. for the kind and helpful support in publishing this work.

References

1. Hannah Ritchie, Max Roser and Pablo Rosado (2022) – "Energy". Published online at OurWorldInData.org. Retrieved from: 'https://ourworldindata.org/energy', last accessed 2023/06/29.
2. Bento, P. M. R., et al.: Ocean wave energy forecasting using optimised deep learning neural networks. Ocean Engineering 219, 108372 (2021).
3. Aly, H. HH.: Intelligent optimized deep learning hybrid models of neuro wavelet, Fourier Series and Recurrent Kalman Filter for tidal currents constitutions forecasting. Ocean Engineering 218, 108254 (2020).
4. Jörges, C., et al.: Prediction and reconstruction of ocean wave heights based on bathymetric data using LSTM neural networks. Ocean Engineering 232, 109046 (2021).
5. Meng, F., et al.: Forecasting tropical cyclones wave height using bidirectional gated recurrent unit. Ocean Engineering 234, 108795 (2021).
6. Amundarain, M., et al.: Neural control for wave power plant during voltage dips. Electric power systems research 92, 96–105 (2012).
7. M'zoughi, F., et al.: ANN-based airflow control for an oscillating water column using surface elevation measurements. Sensors 20.5, 1352 (2020).
8. M'zoughi, F., et al.: Rotational speed control using ANN-based MPPT for OWC based on surface elevation measurements. Applied Sciences 10.24, 8975 (2020).
9. Ghefiri, K., et al.: Multi-layer artificial neural networks based mppt-pitch angle control of a tidal stream generator. Sensors 18.5, 1317 (2018).
10. Bastos, B., et al.: Componentnet: Processing U-and V-components for spatio-Temporal wind speed forecasting. Electric Power Systems Research 192, 106922 (2021).
11. Gan, Z., et al.: Temporal convolutional networks interval prediction model for wind speed forecasting. Electric Power Systems Research 191, 106865 (2021).
12. Liu, X., et al.: Short-term wind power forecasting by stacked recurrent neural networks with parametric sine activation function. Electric Power Systems Research 192 107011 (2021).
13. Toubeau, J-F., et al.: Recalibration of recurrent neural networks for short-term wind power forecasting. Electric Power Systems Research 190, 106639 (2021).
14. Ryu, A., et al.: Battery smoothing control for photovoltaic system using short-term forecast with total sky images. Electric Power Systems Research 190, 106645 (2021).
15. Aly, H. HH.: A proposed intelligent short-term load forecasting hybrid models of ANN, WNN and KF based on clustering techniques for smart grid. Electric Power Systems Research 182, 106191 (2020).
16. Atef, S., et al.: Assessment of stacked unidirectional and bidirectional long short-term memory networks for electricity load forecasting. Electric Power Systems Research 187, 106489 (2020).
17. Chen, H., et al. Day-ahead aggregated load forecasting based on two-terminal sparse coding and deep neural network fusion. Electric Power Systems Research 177, 105987 (2019).
18. Khwaja, A. S., et al.: Joint bagged-boosted artificial neural networks: Using ensemble machine learning to improve short-term electricity load forecasting. Electric Power Systems Research 179, 106080 (2020).
19. Memarzadeh, G., et al.: Short-term electricity load and price forecasting by a new optimal LSTM-NN based prediction algorithm. Electric Power Systems Research 192, 106995 (2021).

20. Sideratos, G., et al.: A novel fuzzy-based ensemble model for load forecasting using hybrid deep neural networks. Electric Power Systems Research 178, 106025 (2020).
21. Wen, Z., et al. Long term electric load forecasting based on TS-type recurrent fuzzy neural network model. Electric Power Systems Research 179, 106106 (2020).
22. Wen, L., et al.: Load demand forecasting of residential buildings using a deep learning model. Electric Power Systems Research 179, 106073 (2020).
23. Zheng, C., et al. A novel RNN based load modelling method with measurement data in active distribution system. Electric Power Systems Research 166, 112–124 (2019).
24. Zhu, K., et al.: A hybrid prediction model based on pattern sequence-based matching method and extreme gradient boosting for holiday load forecasting. Electric Power Systems Research 190, 106841 (2021).
25. Wilberforce, T., et al.: Overview of ocean power technology. Energy 175, 165–181 (2019).
26. NEDO to Conduct World's First 100kW Class Demonstration Test of Ocean Current Power Generation, https://www.nedo.go.jp/english/news/AA5en_100269.html, last accessed 2023/03/13.
27. Shintake, T., et al.: Technical R&D on a Surf Zone WEC. Proceedings of the 12th European Wave and Tidal Energy Conference, Cork, Ireland. Vol. 27. (2017).
28. Fujiwara, R., et al.: Forecasting design values of tidal/ocean power generator in the strait with unidirectional flow by deep learning. Intelligent Systems with Applications 14, 200067 (2022).
29. Kingma, D., et al.: Auto-encoding variational bayes. arXiv preprint arXiv:1312.6114 (2013).
30. Rezende, D., et al.: Stochastic backpropagation and approximate inference in deep generative models. International conference on machine learning. PMLR (2014).
31. Mehmood, N., et al.: Diffuser augmented horizontal axis tidal current turbines. Research Journal of Applied Sciences, Engineering and Technology 4(18), 3522–3532 (2012).
32. Kingma, D., et al.: Adam: A method for stochastic optimization. arXiv preprint arXiv:1412.6980 (2014).
33. Goodfellow, I., et al. Generative adversarial networks. Communications of the ACM 63.11, 139–144 (2020).

Finite Element Analysis of Large-Span Valve Hall Structure Under Extreme Wind Conditions

Dongmei Fan, Wendou Yan, Yi Liang, Weiyuan He, and Gang Lin

1 Introduction

As the core functional area of the converter station, the process layout of the valve hall is complicated. Due to the electrical process restrictions, the valve hall is required to be a barrier-free empty space inside. The clearance height and plane size of the valve hall are both required to be high, which brings great challenges to the architectural structure design. The converter station belongs to the large-span roof structure, the structural vibration caused by wind is obvious, and with the increase of the natural vibration period of the structure, the wind vibration is also enhanced. The effect of wind pressure pulse on wind vibration of structure should be considered and the calculation of wind vibration should be carried out according to random vibration theory.

Studying the structural stability of large-span space grid, Gawronski [1] analyzed the wind load applied in the form of wind force or torque disturbances and obtained similar results. By comparing the Davenport wind spectra and the Kaman wind spectra, Wang et al. used two different wind spectrum calculations of wind-induced response of the structure [2]. Many other researchers studied the seismic performance of similar structures of valve halls on land [3–5] and dynamic responses of similar frame structures under seismic load [6, 7]. At sea, the problem becomes more complex, including seismic, hydrodynamic, and wind load, which need to be considered [8, 9].

D. Fan · W. Yan · Y. Liang · W. He
Guangzhou Power Supply Bureau, Guangdong Power Grid Co., Ltd., Guangzhou, China

G. Lin (✉)
China Energy Engineering Group, Guangdong Electric Power Design Institute Co., Ltd., Guangzhou, China
e-mail: lingang@gedi.com.cn

Z. Peng (ed.), *Trends and Technological Challenges in Green Energy*, Green Energy and Technology, https://doi.org/10.1007/978-3-031-55579-4_7

In this chapter, Davenport wind spectra were transformed into pulsating wind force. Firstly, the finite element model of the valve was established, and the wind vibration response of the valve was obtained through time history analysis. The displacement and stress changes of each node of the grid structure were discussed, and the key nodes and elements were selected for the characterization. Furthermore, the stress and deformation of the large-span valve hall grid structure under the action of a strong typhoon were calculated to ensure the safety and stability of the structure under extreme environments.

2 Numerical Investigation

2.1 *Problem Statement*

The cross section of the large-span valve hall structure is shown in Fig. 1. The grid support is arranged on the lower chord layer, and the support is evenly arranged along all sides. The headroom of the present valve hall is 61 m wide and 73 m long, and the standard height of lower chord bracing is 20 m. The valve hall grid is rectangular layout, and the vertical quadrangular double layer net is adopted, the grid size being 4 m × 4 m. The grid adopts variable height design, and the middle of the valve hall is about 24.7 m.

The finite element model of the large-span valve hall structure is shown in Fig. 2. Both the top chord and the web member mainly bear axial load, and bar element LINK180 is used for these parts. While lower chord needs to bear transverse bending moment, and beam element BEAM188 is employed for lower chord parts. An ideal elastoplastic material model is used, and the elastic modulus and Poisson's ratio are 206 GPa and 0.3, respectively. The density of the material is 7850 kg/m^3, and the yielding stress is 355 MPa. The whole finite element model has 3828 bar element

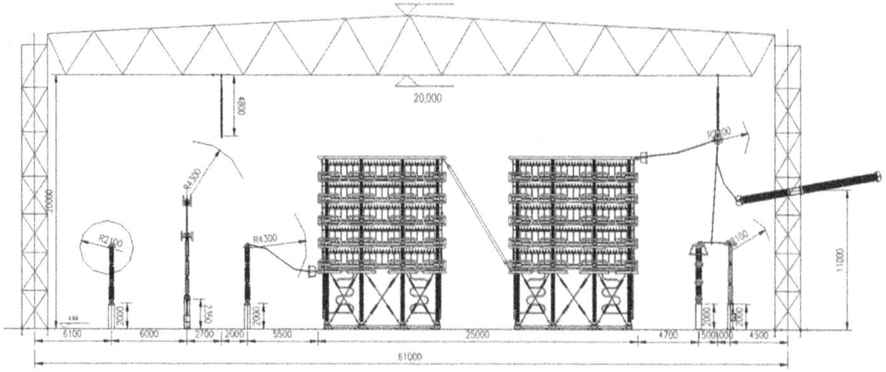

Fig. 1 Cross section of the large-span valve hall structure

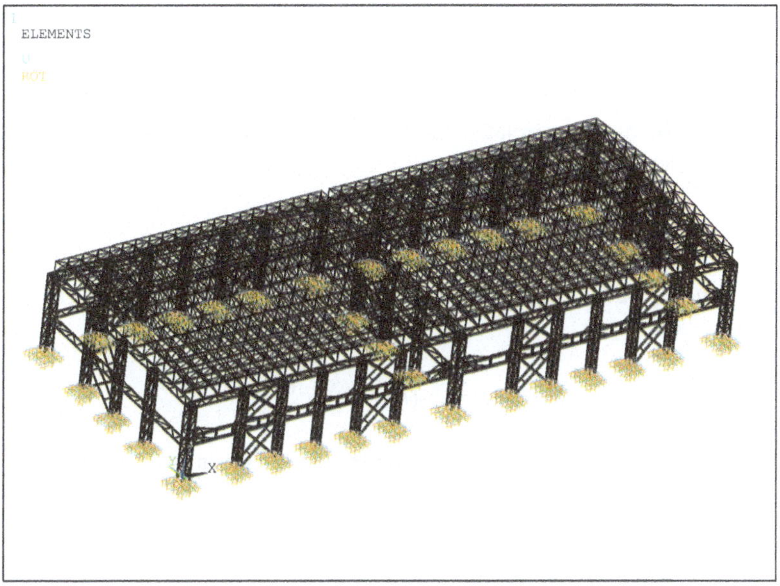

Fig. 2 Finite element model of the large-span valve hall structure

and 5534 beam elements, and the total node number is 3149. The bottom of the large-span valve hall structure is clamped during the computation.

2.2 *Wind Load*

The Davenport wind speed spectrum is employed in the present investigation [10]:

$$\frac{(nS_v(n))}{K\overline{v}_{10}^2} = \frac{4x^2}{(1+x^2)^{4/3}} \quad (n \geq 0) \tag{1}$$

where n denotes the frequency, $\overline{v}_{10}$ is the average wind velocity at 10 m height during 10 min, K is the ground roughness coefficient, which varies between 0.003 and 0.03, and $x = 1200n/\overline{v}_{10}$.

In the finite element analysis of structural dynamics, random wind field is adopted, and the power spectrum is set to be the same with the Davenport spectrum. Random wind field is assumed to be a stationary Gaussian random process and is constructed with harmonic superposition method through a series of harmonic functions with random amplitudes and frequencies. The random wind velocity is written as [11–13]

$$\boldsymbol{v}(\boldsymbol{x},t)=\sum_{n=1}^{N}\left[\boldsymbol{A}_n\cos(\boldsymbol{k}_n\cdot\boldsymbol{x}+\omega_n t)+\boldsymbol{B}_n\sin(\boldsymbol{k}_n\cdot\boldsymbol{x}+\omega_n t)\right] \tag{2}$$

where $\boldsymbol{A}_n$ and $\boldsymbol{B}_n$ are uncorrelated random vectors and are chosen according to a prescribed power law energy spectrum:

$$|\boldsymbol{A}_n|^2=|\boldsymbol{B}_n|^2=2E(k_n)\Delta k_n \tag{3}$$

where $\boldsymbol{k}_n$ denotes wave vectors, and the number vector of wind velocity can be given by

$$k_n=|\boldsymbol{k}_n|=k_1(k_N/k_1)^{(n-1)/(N-1)} \tag{4}$$

N denotes the node numbers, and ω_n is the frequency given by

$$\omega_n=\lambda\sqrt{k_n^3E(k_n)} \tag{5}$$

where λ is the instability parameter, which varies between 0 and 1, and 0.5 is adopted in the present investigation.

Power spectrum is assumed to have the expression

$$E(k_n)=E_0L(k_nL)^{-p} \tag{6}$$

where the upper limit length scale of the inertial range $L=2\pi/k_1$, E_0 is a constant, and p is the power of the energy spectrum.

According to the aforementioned algorithm, random wind velocity can be constructed. The total time and time interval are 600 s and 0.1 s, respectively. The ground surface roughness = 0.005, and the node number is 6000. The frequency interval is 0.001–6 Hz.

The previous method shows the variation of wind velocity with time, while the variation of wind velocity with space can be determined according to the following steps. The relationship between the wind velocity and wind pressure is given by

$$p-p_0=W_PC_Pd^2 \tag{7}$$

where p and p_0 denote average wind pressure and reference wind pressure, respectively, W_P is standard wind pressure, and d denotes the length scale of the upwind surface. The shape factor of wind load C_P for double-sloped roof is given in Fig. 3.

Then, the structural dynamic equations can be written as

$$\boldsymbol{M}\ddot{\boldsymbol{U}}+\boldsymbol{C}\dot{\boldsymbol{U}}+\boldsymbol{K}\boldsymbol{U}=\boldsymbol{F} \tag{8}$$

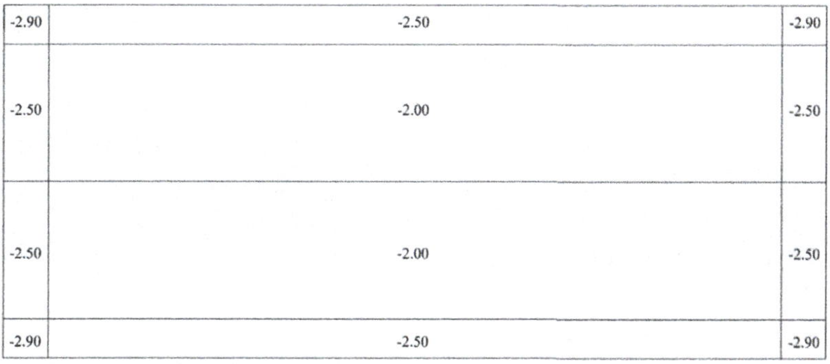

Fig. 3 Shape factor of wind load for double-sloped roof

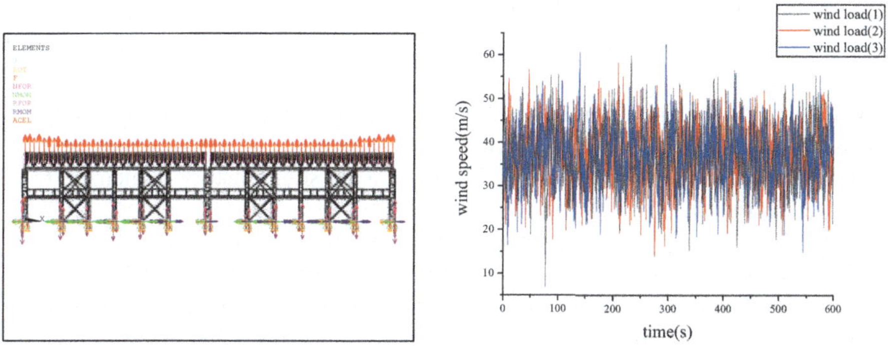

Fig. 4 Illustration of wind pressure on the finite element model

where $\boldsymbol{M}$, $\boldsymbol{C}$, and $\boldsymbol{K}$ denote mass matrix, damping matrix, and stiffness matrix, respectively, $\ddot{\boldsymbol{U}}$, $\dot{\boldsymbol{U}}$, and $\boldsymbol{U}$ denote acceleration vector, velocity vector, and displacement vector, respectively, and $\boldsymbol{F}$ is the external force vector. The finite element equations of equation (8) can be solved with proper boundary conditions and initial conditions. The illustration of wind pressure on the finite element model is given in Fig. 4.

2.3 Numerical Results

In order to perform a safety assessment of the structure under extreme wind conditions, finite element analysis is conducted with the model described in Sects. 2.1 and 2.2 under an average wind speed of 37 m/s, which is supposed to be a typical extreme condition of hundred-year weather. Transient structural dynamic analysis is

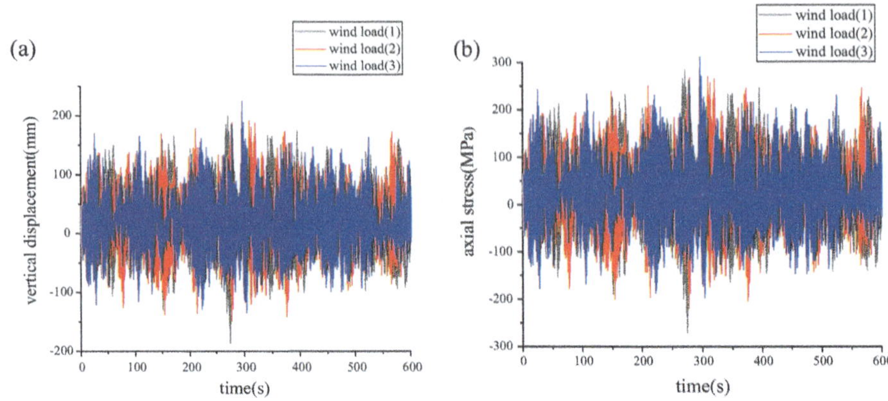

Fig. 5 Typical response of the large-span valve hall structure: (**a**) vertical displacement history of node 2011; (**b**) axial stress history of element 6063

performed with the finite element method. Node 2011 and element 6063 are selected as the key components to evaluate the health condition of the structure, and the vertical displacement history of node 2011 and axial stress history of element 6063 are given in Fig. 5.

It can be seen that the maximal wind velocities for the three cases are 58.3 m/s, 59.8 m/s, and 62.5 m/s, respectively; accordingly, the maximal vertical displacement and maximal axial stress under these three cases are 200.4 mm and 284.8 MPa, 194.3 mm and 273.1 MPa, and 226.9 mm and 314.9 MPa, respectively. Under the extreme condition of average 37 m/s wind velocity, both the maximal vertical displacement and maximal axial stress are smaller than the corresponding allowable values.

In order to further understand the bearing limit of the structure under extreme wind load, the dynamic response of the structure under wind speeds of 40 m/s, 45 m/s, and 50 m/s was calculated. The results are shown in Fig. 6.

It can be seen from Table 1 that, under the average wind load of 40 m/s, the axial stress of some components at a certain time is very close to the yield limit of 355 MPa. There is a certain safety risk at this time, and the valve hall structure should be considered for reinforcement treatment. If the wind speed continues to increase by 5 m/s, the maximum axial stress has exceeded the yield limit, and local instability failure may occur. At the same time, it can also be noted that the increase in stress caused by wind speed from 45 m/s to 50 m/s is significantly higher than that from 40 m/s to 45 m/s. At the wind speed of 50 m/s, the maximum axial stress reaches 570 MPa, which is 1.6 times the yield limit, and the structure is in great danger.

In the example of wind speed of 50 m/s, the most dangerous moment occurs at 396.1 s. The equivalent stress contour of the grid structure and latticed column structure at 396.1 s are given in Fig. 7.

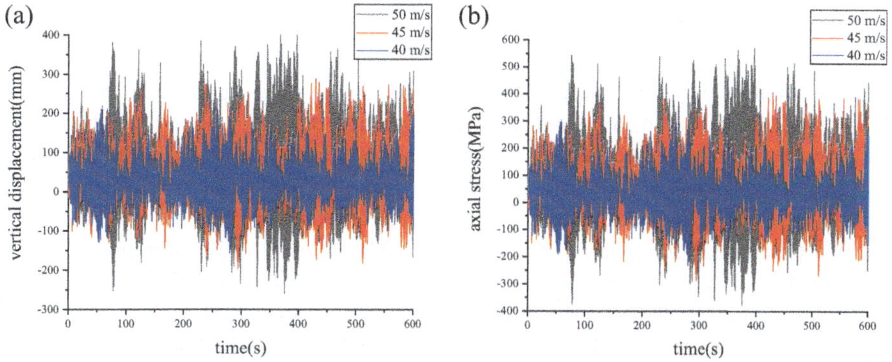

Fig. 6 Typical response under extreme wind loads: (**a**) vertical displacement history of node 2011; (**b**) axial stress history of element 6063

Table 1 Dynamic response of the structure under extreme wind loads

Average wind speed (m/s)	Maximum instantaneous wind speed (m·s^{-1})	Maximum vertical displacement (mm)	Maximum axial stress (MPa)
40	63.1	209.5	338.3
45	69.6	291.1	411.2
50	79.8	400.7	570.8

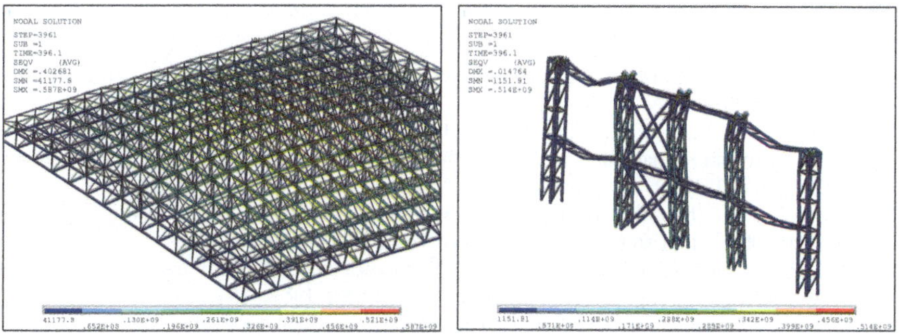

Fig. 7 Equivalent stress contour of the grid structure and latticed column structure

It can be seen that, under the average wind velocity of 50 m/s, axial stress of some bar elements of the structure has already exceeded the yielding stress of the material, which is 355 MPa. Plastic deformation occurs at some parts of the large-span valve hall structure and raises the risk of structural failure.

3 Conclusions

In this chapter, the finite element solid model of the large-span valve hall of the converter station was established first. The Davenport wind velocity spectrum was introduced to simulate the random time domain of the natural wind. Combined with the shape coefficient of the double-slope structure, the multidimensional spatial and temporal simulation of the wind load of the valve hall was conducted. In the case of 37 m/s (the maximum wind speed in the local 100 years), the maximum vertical displacement is 186.4 mm and appears in the center of the grid. This is still 23% richer than the maximum allowable mid-span deflection of 244.0 mm. The maximum axial stress also appears in the same position of 284.9 MPa, which is less than the material yield limit of 355 MPa, and the richness is about 20%. As the wind load continues to increase, the axial stress of the member at a certain time reaches 284.9 MPa under the wind speed of 40 m/s, which is very close to the yield limit of 355 MPa. If the wind load continues to increase, the structure may no longer be safe. To sum up, under the meteorological environment here, the wind resistance performance of the large-span grid structure of this project is sufficient.

Acknowledgment This project was supported by the Southern Power Grid Company [Project No. GZHKJXM20210046 (080041KK52210005)].

References

1. Gawronski W.: Modeling wind-gust disturbances for the analysis of antenna pointing accuracy. IEEE Antenn. Propag. M. 46 50–58 (2004)
2. Wang L., Qu W. L., Ji B. F., Wang Y. F. and Chen J. P.: Comparative analysis of Davenport wind spectrum and Kaman wind spectrum in the towering structural design. Appl. Mech. Mater. 501 823–826 (2014)
3. Yang Z., Xie Q., Zhou Y. and Mosalam K. M.: Seismic performance and restraint system of suspended 800 kV thyristor valve. Eng. Struct. 169 179–187 (2019)
4. He C., Xie Q., Yang Z. and Xue S.: Seismic performance evaluation and improvement of ultra-high voltage wall bushing-valve hall system. J. Constr. Steel Res. 154 123–133 (2019)
5. Enblom R., Coad J. N. O. and Berggren S.: Design of HVDC converter station equipment subject to severe seismic performance requirements. IEEE T. Power Deliver. 8 1766–1772 (2002)
6. Heru P., Mulia O., Sjahril A. R. and Reza A. K.: Evaluation of notional loads magnitude to three calibration moment resisting frames subjected to Indonesian seismic load. Int. J. Mech. Eng. Rob. Res. 9 35–41 (2020)
7. Rani K. S. and Reddy V. K.: Design optimization of the tubing assembly structure for seismic conditions. Int. J. Mech. Eng. Rob. Res. 3 542–550 (2014)
8. Sun Z., Zhao S., Bi C., Chen Q., Huang S. and Chen J.: Dynamic response analysis of an offshore converter platform with valve towers under seismic excitation. Symmetry. 14 1–15 (2022)
9. Sun Z., Yu Y., Wang H., Huang S. and Chen J.: Dynamic response analysis of offshore converter station based on vector form intrinsic finite element (VFIFE) method. J. Mar. Sci. Eng. 10 1–21 (2022)

10. Davenport A. G.: The spectrum of horizontal gustiness near the ground in high winds. Q. J. R. Meteorol Soc. 87 194–211 (1961)
11. Flohr P. and Vassilicos J. C.: A scalar subgrid model with flow structure for large-eddy simulations of scalar variances. J. Fluid Mech. 407 315–349 (2000)
12. Nicolleau F. and Yu G.: Two-particle diffusion and locality assumption. Phys. Fluids 16 2309–2321 (2004)
13. Kraichnan R. H.: Diffusion by a random velocity field. Phys. Fluids 13 22–31 (1970)

Part III
Renewable Energy Systems and Thermal Engineering

Dynamics of Renewable Energy Penetration for the Malaysian Electricity Generation

Zong Qing Yeoh, Andrew Huey Ping Tan, Anwar P. P. Abdul Majeed, Wei Chen, and Eng Hwa Yap

1 Background and Motivation

Primary energy demand and energy-sourced carbon dioxide emissions have increased by 5.8% and 5.9%, respectively [1]. Despite observing a slight decrease, fossil fuels remain the highest primary energy supply in 2021, at above 80%, comparable to 2015 levels [1]. On a national level, primary energy consumption in Malaysia rose by 2.4% in 2021, with an average of 2.1% in the 2011–2021 period [1]. At the same time, renewable energy (RE) generation grew by 12.3% in Malaysia in 2021, 2.7% shy of the global growth of 15% in the same year [1].

The reserves of nonrenewable resources (NRRs) such as fossil fuels, natural gases, and coal will not last more than 50 years if NRRs are continually used [2]. In order to retard the depletion of NRRs, nations worldwide, including Malaysia, should consider expanding the generation and usage of RE.

Malaysia, a Southeast Asian equatorial country, is divided into two separate areas, namely Peninsular Malaysia and East Malaysia. Owing to its tropical climate of being generally hot and humid with an average temperature of 27 °C throughout the year with an average rainfall of 250 cm a year [3], Malaysia has great potential for the development of RE such as solar energy, hydropower, biomass energy, and wind energy. Besides, the government of Malaysia has moved from the four-fuel policy (oil, natural gas, coal, and hydropower) to five-fuel policy (oil, natural gas,

Z. Q. Yeoh
UOW Malaysia KDU Penang University College, Georgetown Penang, Malaysia

A. H. P. Tan (✉) · W. Chen
School of Intelligent Manufacturing Ecosystem, Xi'an Jiaotong-Liverpool University, Suzhou, China
e-mail: andrew.tan@xjtlu.edu.cn

A. P. P. A. Majeed · E. H. Yap
School of Robotics, Xi'an Jiaotong-Liverpool University, Suzhou, China

Z. Peng (ed.), *Trends and Technological Challenges in Green Energy*, Green Energy and Technology, https://doi.org/10.1007/978-3-031-55579-4_8

coal, hydropower, and renewable energy) under the 8th Malaysia plan since 2001 [4]. In the same year, the small renewable energy power (SREP) program was launched by the Malaysia government to promote the small-scale renewable electricity in Malaysia but it ended in 2010 [5].

In order to achieve a healthy penetration of RE in Malaysia, the best combination of RE penetration and share must be investigated. This is important because, like many other countries, to achieve the maximum reliability of RE generation and usage, the best mix of RE penetration and share needs to be suitable according to Malaysian geographical conditions. Besides, adoption and expansion of RE technologies are similar in complexity with other technologies, in ways of facing complexities such as investment risks, technological readiness level, socioeconomic conditions, and governmental policies. Variables within these complexities will also exhibit dynamic change over time, which at times might render the respective original goals to be changed as well. As a result, this makes the adoption and expansion of RE technologies in a country such as Malaysia a complex problem involving intertwined crucial variables [6].

As a response to this problem, this chapter will attempt to investigate the scenario of RE penetration into Malaysia by using a system dynamics approach. By identifying the key indicators involved in the Malaysian RE arena, their interrelationships can be explored and quantified before being system dynamically simulated in order to predict their behaviors over time. As a result, this chapter's system dynamics model and simulation will provide a systems understanding of the RE penetration and share behavior in Malaysia, before providing a data-backed policy recommendation for the country.

2 Literature Review

2.1 Renewable Energy and Its Potential in Malaysia

Renewable energy is an energy category that has electricity conversion processes that do not involve the consumption of exhaustible finite resources such as fossil fuels and uranium. Generally, RE resources include sunlight, wind, water, biomass materials, and earth's geothermal heat. RE plays a critical role in reducing greenhouse gases (GHG) and CO_2e emissions. Despite RE having high growth rates each year, it is still only a small portion of the global energy share.

Table 1 shows the potential RE that can be expanded in Malaysia [7]. The potential of biomass is about 1300 MW that is technically feasible. The largest source of biomass is oil palm as it is one of the main exports of Malaysia. About 4.49 million hectares of land in Malaysia, 13.57% of the total land area, were used for oil palm plantation in 2016 [7]. Oil palm wastes can be used for electricity production. Additionally, municipal solid waste (MSW) produced by Malaysian residents has the potential for 400 MW of capacity. MSW can produce biogas

Table 1 Renewable energy potential in Malaysia [7]

Renewable energy	Potential (MW)
Hydropower	22,000
Mini-hydro	500
Biomass/biogas	1300
Municipal solid waste	400
Solar PV	6500

through the process of anaerobic, combustion, and other processes to produce energy.

For water-sourced energies, the potential for hydropower far exceeds that of mini-hydro, at 22,000 MW and 500 MW, respectively. Out of this, the installed hydropower generation in Malaysia is about 6275 MW [8]. While mini-hydro does not have as high a power potential as hydropower, it is a convenient power source for off-the-grid and isolated places.

Solar PV has the potential of 6500 MW that can be built in Malaysia. However, there are only 140 MW of installed solar capacity. The reason for this large potential is simply due to Malaysia's geographical location, which receives long hours of sunshine 365 days a year, ranging from 4.21 kWh/m^2 to 5.56 kWh/m^2 a day [9]. The highest solar radiation was estimated at 6.8 kWh/m^2 in August and November while the lowest was 0.61 kWh/m^2 in December [10].

While wind energy is popular in countries such as the United States, China, Germany, and many others, it has limited potential in Malaysia such as Pulau Perhentian [11]. The mean wind speed is considerably low for the operation of a wind turbine. Mountain areas have relatively stronger winds, but the cost of building a wind turbine in these areas will offset any benefit gained from building it in the first place.

2.2 *Electricity Generation and Consumption in Malaysia*

According to Malaysia's Energy Commission, the electricity generation in Malaysia increases every year [12]. Hydropower has also been steadily increasing since 2000. Hydropower plays an important role in electricity generation in Malaysia, contributing to 12.78% of total electricity generation.

A thermal station is a conventional power station that converts heat energy into electricity using typical NRR as raw materials. Basically, thermal stations work on the principle of generating enough heat to convert water into steam in order to spin the turbine and drive the electric generator, in alignment with the basics of the Rankine Cycle. The water will condense back to liquid form after being cooled to complete the cycle. The main sources to operate the thermal station are coal and coke, which will produce sufficient heat to convert heat energy into electricity energy. In Malaysia, thermal station is the largest contributor to total electricity generation, which is about 83% [12].

Co-gen or cogeneration is a concept of electricity generation where electricity is jointly generated alongside useful heat. This way, co-gen power stations are more efficient compared to conventional thermal stations. Co-gen offers more energy savings in the range of 15–40% compared to traditional power stations [12]. In Malaysia, co-gen power station produces about 3.97% of the total national electricity produced [12].

2.3 Resources Used for Electricity Generation

There are several resources that are used to produce electricity in Malaysia, but the resources that are mostly used for electricity generation are natural gas and crude oil [13]. The highest percentage, at 68.2%, across all resources that are used to supply the electricity in Malaysia comes from natural gas [13]. The second highest is crude oil, which is about 28.1% of overall resources supplied [13]. Together, they make up the lion's share of the resources used for electricity generation in Malaysia. Apart from that, Malaysia does import resources for the same reason, mainly coal (31.7% of total resources imported) and oil products (30% of total resources imported), apart from natural gas and crude oil as well.

3 Methodology

3.1 System Dynamics Process

Figure 1 shows the system dynamics method, as adapted from Sterman [14], employed in this study. The process is started by identifying particular subproblems within the boundary of this research, namely the identification of key indicators and interrelationships between them. A hypothesis is then formed on how these identified variables affect one another. The third step involves constructing causal loop and stock flow diagrams, before which they can be simulated using computer software. The equations that govern their relationships are also mathematically

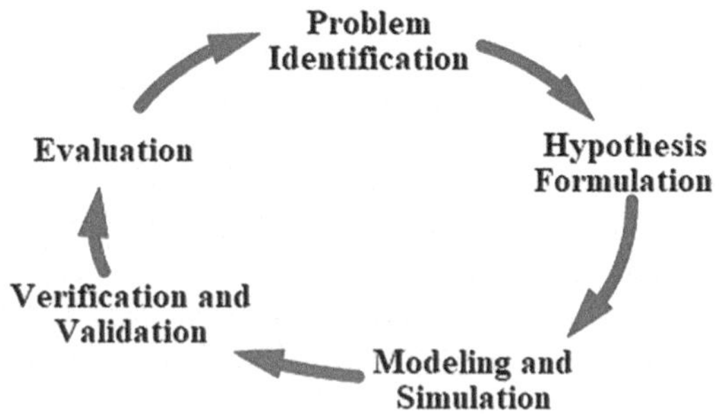

Fig. 1 The system dynamics process

keyed-in in this step. Upon obtaining results, they are then verified against scientific principles and validated against published data. Finally, these results will be evaluated before a conclusion and policy recommendation are put forward.

3.2 Concepts of System Dynamics Modeling

Causal Loop Diagrams

System dynamic models start out with causal loop diagrams (CLDs). CLDs build on the core concepts of feedback, where identified variables are linked to each other and investigated if one variable will cause an increase or decrease in the other, hence causality. There are two types of CLD, namely reinforcing loop and balancing loop.

Figure 2 shows the example of a reinforcing loop and a balancing loop. In the first loop, there is money in the bank and interest from the bank due to this deposited money. These two variables are linked to each other and form a reinforcing loop because the more money there is in the bank, the more interest is obtained from the bank. Also, the more interest that is obtained from the bank, the more money will be added to the bank. If such a loop is left alone without interference from external factors or other subsystems, the values will likely grow exponentially out of control over a long period of time.

The second part of Fig. 2 shows a balancing loop example, given by interlinked variables of "feeling of hunger" and "eating of food." If the feeling of hunger were to be increased, the activity of eating food will be increased as well. However, once the food has been successfully eaten, the feeling of hunger would be reduced. As the feeling of hunger has been reduced, it would naturally reduce the eating of food as well. However, with a reduction in eating food (and given enough time), the feeling of hunger will rise again.

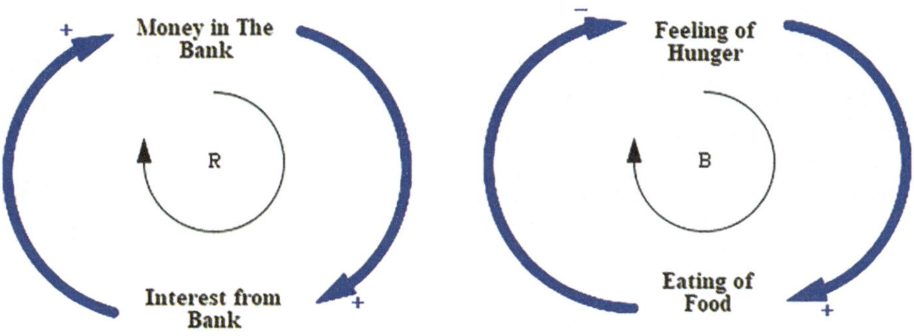

Fig. 2 Causal loop examples

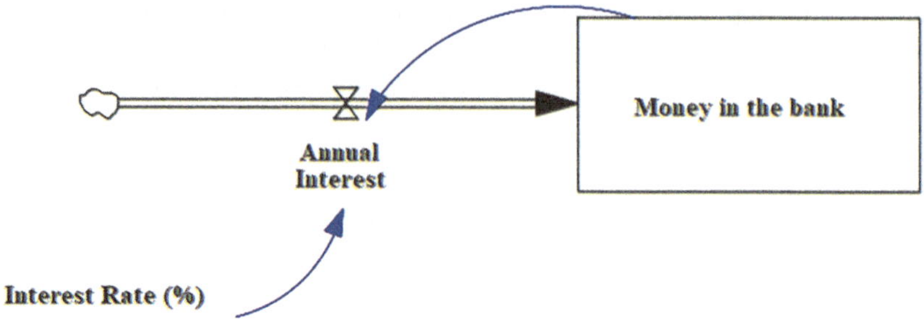

Fig. 3 Example of a stock and flow diagram

Stock and Flow Diagram

Stock and flow diagram (SFD) is the simulatable version of the constructed CLDs. When the first part of Fig. 2 is converted into an SFD, it would be similar to Fig. 3.

As it is simulatable, each variable in Fig. 3 would either have a value or an equation attached to it. For example,

$$\text{Annual Interest} = \text{Money in the bank} * \frac{\text{Interest Rate}}{100} \$/\text{Year} \tag{1}$$

$$\text{Interest rate} = 4\% \tag{2}$$

$$\text{Money in the bank} = \int_0^t \text{Annual Interest dt } \$ \tag{3}$$

4 System Models, Simulations, Results, and Discussion

4.1 Population, GDP, and Energy Consumption in Malaysia

Figure 4 shows the model subsystem of Malaysia's population, GDP, total energy consumption, GDP per capita, and energy per capita. The population will grow in the next 50 years from approximately 31.1 million people in 2016 to 70.3 million in 2066, as shown in Fig. 5.

The GDP per capita in Malaysia will grow exponentially in the next 50 years from 11028.2 USD in 2016 to 35886.1 USD in 2066. Similarly, the GDP per capita will grow exponentially as well from 4.6 million Wh in 2016 to 12.8 million Wh in 2066.

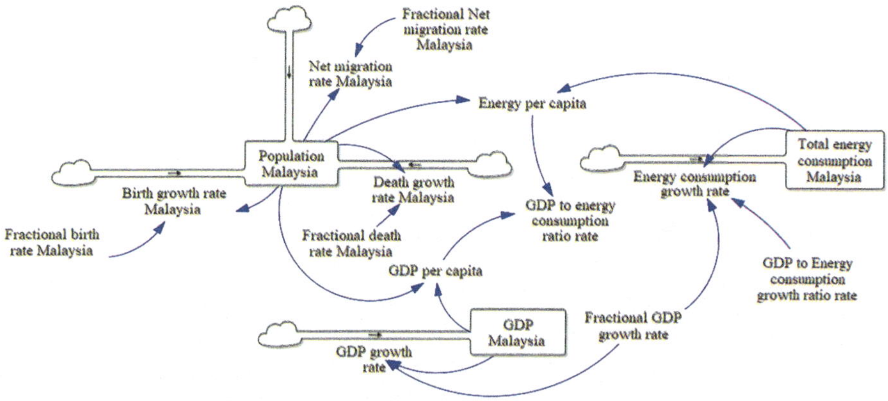

Fig. 4 Population, GDP, and energy consumption in Malaysia

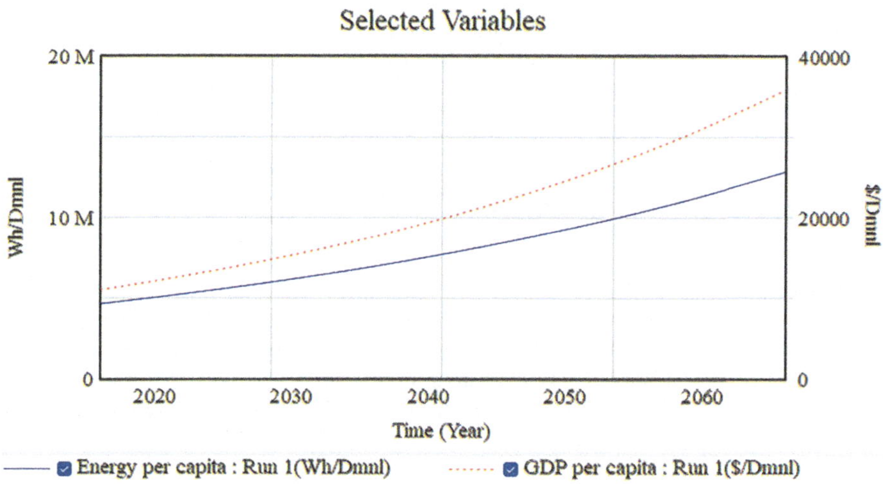

Fig. 5 Energy and GDP per capita

4.2 CO_2 Emissions in Malaysia

Figure 6 shows the model of CO_2e emissions in Malaysia from nonrenewable, renewables, and daily human activities and its projections to 2066. While there may be other sources of CO_2e emissions, the sum total of CO_2e emissions within this study's context is taken from the total of these three sources, given by "Total CO2e emissions." An average human exhales between 0.90 and 1.04 kg a day, which is approximately 380.73 kg a year. Based on the assumption of exponential growth of population and the corresponding energy capacities to support the population, the

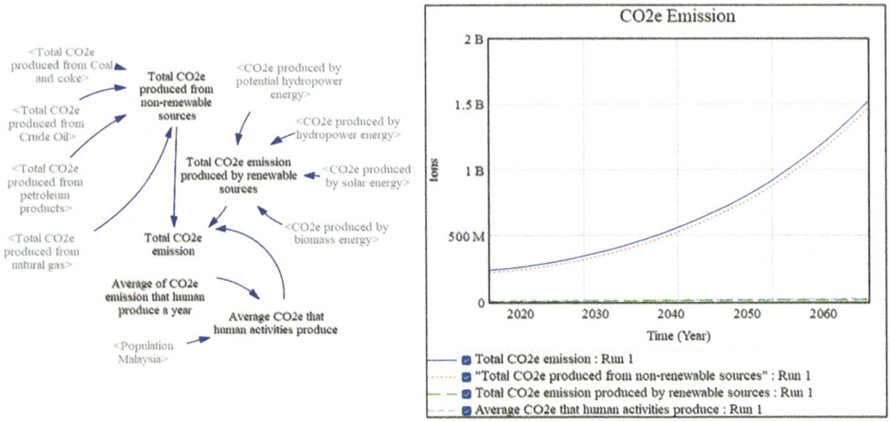

Fig. 6 CO_2e emissions in Malaysia

CO_2e emissions exhibit an exponential growth as well, growing from 293 million tons in 2023 to 1.5 billion tons in 2026. Most of these emissions are contributed by nonrenewable energy capacities.

4.3 Forested Areas in Malaysia

Figure 7 shows the interrelationships between forested areas in Malaysia, CO_2e emissions, and equivalent trees needed to offset these emissions. The estimated typical area needed to plant a tree is 5.06 m^2, but in reality there are more trees in the forested area of Malaysia. By assuming an average CO_2e absorption per year for a unit tree, the total trees needed to neutralize the CO_2e emission released can be calculated by

$$\text{total trees} = \frac{\text{Total } CO_2\text{e emission}}{\text{Average of a tree } CO_2\text{e absorption rate per year}} \tag{4}$$

Figure 7 also shows the projected forested land use in Peninsular, Sabah, Sarawak, and the trees needed to neutralize the CO_2e emissions. In an ideal case, where equivalent number of trees are planted for every ton of CO_2e emitted, the ratio of neutralizing trees' land use to CO_2e emitted is seen to be very high from above 30 in the 2020s to above 200s in 2066. This is in tandem with the assumed exponential growth of national GDP under a very ideal case.

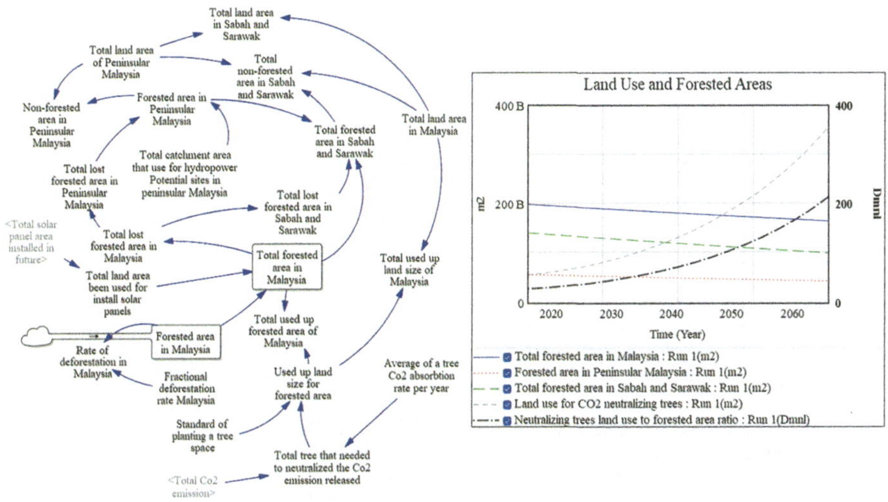

Fig. 7 Forested area and CO_2 absorption in Malaysia

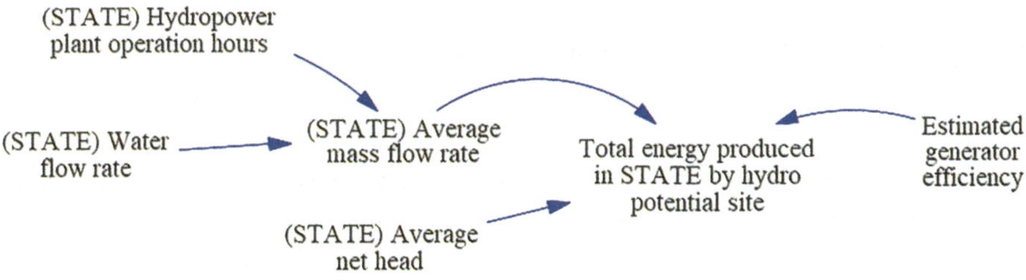

Fig. 8 Micro-hydropower sites in Malaysia by state

4.4 Micro-Hydro Potential Sites by State

Figure 8 shows the subsystem to calculate the technically feasible micro-hydro energy produced by each state in Malaysia. In this subsystem, the sum of all hydropower and micro-hydropower potential sites will be added for the states. The energy produced at each site will be calculated using water flow rate and net head data as taken from Raman et al. This calculation is also made by assuming the hydropower site operates for 24 h a day.

Based on Fig. 9, Perak has the highest micro-hydropower sites in peninsular Malaysia which have 34 potential micro-hydropower sites that estimate will produce 9944 kW with the actual mass flow rate of 12,171 liter/s, with an ideal flow rate of 14,560 liter/s. Penang, Perlis, and Malacca are the states in peninsular Malaysia that have no potential for micro-hydropower sites. The states in peninsular Malaysia that have the lowest potential for micro-hydropower sites are Selangor, which have two

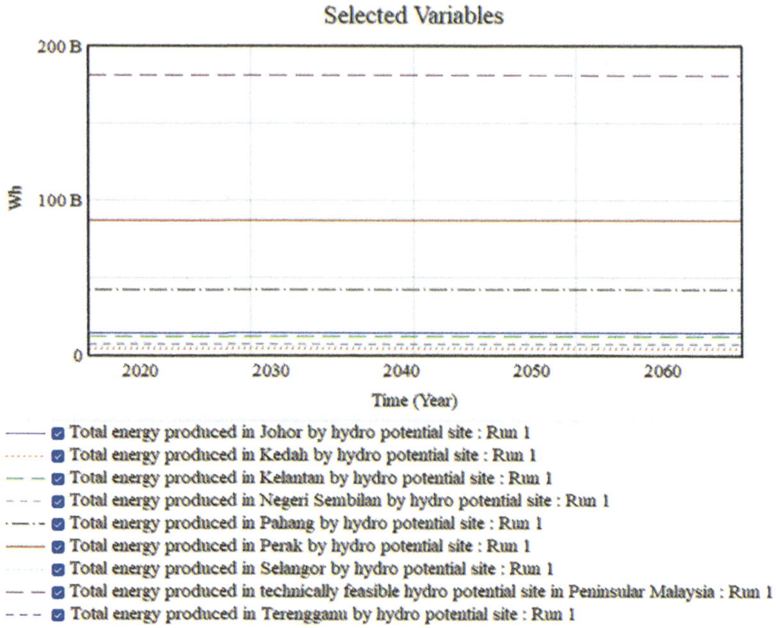

Fig. 9 Total hydro energy potential in Malaysian states

sites that estimate will produce 343 kW. The total potential for micro-hydropower in peninsular Malaysia is estimated to be at 20,000 kW or about 1.75 MWh worth of energy per year.

4.5 Energy Production in Malaysia

Figure 10 shows the model subsystem of various energy types and production in Malaysia, alongside the share of renewables and nonrenewables while Fig. 11 shows their predicted behaviors from 2020s all through to 2066. As penetration of renewable energy would be a variable decided independently, nonrenewable energy share in this case would thus be dependent on renewable energy share. Renewable energies, suitable for Malaysia, are the sum total of solar power, hydropower, biogas, biomass, and biodiesel power energy. From the simulation, it is observed total energy production increased from 156 TWh in 2016 to 971 TWh in 2066. Renewable energy produced grows steadily over this 50-year period, from 21 TWh in 2016 to 106 TWh in 2066. While this may be encouraging on its own, but renewables growth pales in comparison to nonrenewables growth, which is observed to exhibit an exponential growth all the way from 2016 to 2066. Renewables share is seen to have decent growth from the mid to late 2020s, and peak at 22% (with

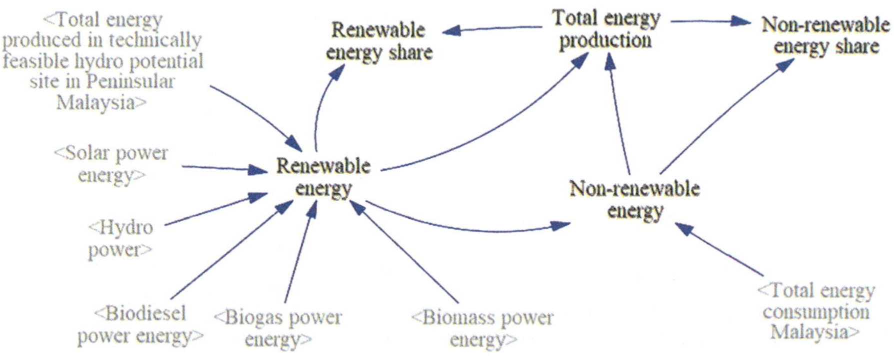

Fig. 10 Subsystem of energy types, share, and production in Malaysia

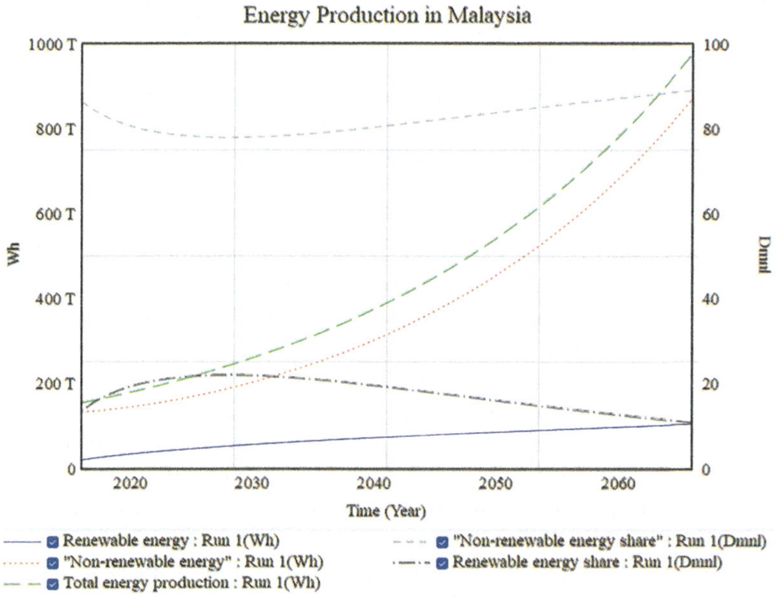

Fig. 11 Behavior of energy types, share, and production in Malaysia

nonrenewables share at 78%) at about 2028, before starting to decrease again. This behavior is aligned with common knowledge, considering that the main renewable sources in Malaysia are solar and hydro and there is a limited amount of land suitable for solar farm and finite amount of rivers and dams for hydropower. On the contrary, despite the nonrenewable resources, such as coal, natural gas, and oil, having a limited amount too, their global resource and reserves are far larger than the natural renewable limits in Malaysia. Consequently, within the context of this study's boundaries, the exponential growth of energy demand in this country does not stop the exponential growth of nonrenewable capacities.

5 Conclusions

The penetration of expansion of RE into a country such as Malaysia is indeed a complex system, consisting of numerous different variables at play. Variables from various different dimensions, such as the general population, socioeconomics, technology availability, geographical suitability, environmental emissions, and policies, affect the system at various points of concern. Through the system dynamics simulation, insights into how to best approach the penetration of RE into Malaysia can be derived.

Firstly, the energy demands of the population albeit being constant or just increasing little yearly, the exponential growth of population would result in the rapid growth of energy demand as well. By accepting this fact, efforts to continually expand on energy technologies and increasing electricity generation capacities must therefore be maintained or increased. This growth is also largely influenced by the growth of national GDP, which encourages more activities, and hence more energy needed.

Secondly, neutralizing the emissions of CO_2e emitted with planting more trees may be a good effort, but it would not be practical, considering the exponential growth of demand, and hence its emissions. This is due to the scarcity of available land, which is limited by the finite amount in the country. While planting trees is a step in the right direction in controlling CO_2e emissions, it must be combined with other efforts in order to maximize its benefits.

Thirdly, there is a limit to RE share penetration in the country, which is seen to peak at 22% in 2028, with a corresponding non-RE share of 78%. This is due to the combined factors of current technological levels, geographical limits in the country conducive to RE technologies, and the assumed exponential growth. On top of that, this scenario may not be sufficient to retard the CO_2e emissions from other sources, as shown in Fig. 6, where CO_2e emissions pretty much follow the same exponential growth pattern as population and GDP.

While this study shows that little can be done on the RE penetration and the impending exponential growth of demands, it is still just within the context of the selected boundaries of this study. Possible technological improvement in the future coupled with a more efficient way of living, among many other factors, could perhaps provide a more ideal condition for healthy RE penetration and much less CO_2e emissions.

References

1. BP (2022) BP Statistical Review of World Energy 2022,(71st edition). Bp 1–60
2. Suslick SB, Machado IF (2002) Nonrenewable Resources
3. Shaikh PH, Nor NBM, Sahito AA, et al (2017) Building energy for sustainable development in Malaysia: A review. Renew Sustain Energy Rev 75:1392–1403. https://doi.org/10.1016/j.rser.2016.11.128
4. Maulud AL, Saidi H (2012) The Malaysian Fifth Fuel Policy: Re-strategising the Malaysian Renewable Energy Initiatives. Energy Policy 48:88–92. https://doi.org/10.1016/j.enpol.2012.06.023
5. Sovacool BK, Drupady IM (2011) Examining the small renewable energy power (SREP) program in Malaysia. Energy Policy 39:7244–7256. https://doi.org/10.1016/j.enpol.2011.08.045
6. Frensch PA, Funke J (2014) Complex Problem Solving The European Persepctive. Psychology Press, New York
7. Mekhilef S, Saidur R, Safari A, Mustaffa WESB (2011) Biomass energy in Malaysia: Current state and prospects. Renew Sustain Energy Rev 15:3360–3370. https://doi.org/10.1016/j.rser.2011.04.016
8. IHA (2022) Hydropower Status Report 2022. Int Hydropower Assoc 1–83
9. Azhari AW, Sopian K, Zaharim A, Al Ghoul M (2008) A new approach for predicting solar radiation in tropical environment using satellite images - Case study of Malaysia. WSEAS Trans Environ Dev 4:373–378
10. Ong HC, Mahlia TMI, Masjuki HH (2011) A review on energy scenario and sustainable energy in Malaysia. Renew Sustain Energy Rev 15:639–647. https://doi.org/10.1016/j.rser.2010.09.043
11. Energydata.info Global Wind Atlas. https://globalwindatlas.info/en/area/Malaysia
12. Commision E (2019) Annual Report 2019 (Laporan Tahunan). 17–18
13. Chong C, Ni W, Ma L, et al (2015) The use of energy in Malaysia: Tracing energy flows from primary source to end use. Energies 8:2828–2866. https://doi.org/10.3390/en8042828
14. Sterman JD (2000) Business Dynamics: Systems Thinking and Modeling for a Complex World. McGraw-Hill

Optimal Sizing and Operation of Hybrid Renewable Energy System: A Techno-Economic Analysis of Residential Building Considering Seasonal Storage

Kaichen Qu and Hong Zhang

1 Introduction

Renewable energy resources, such as solar, wind [1], and geothermal [2], have received considerable attention in recent years owing to significant advancements in energy structure transformation, environmental impact reduction, and economic profits [3]. However, renewable energy generation is inherently susceptible to power mismatches due to the uncontrollable and intermittent nature of natural resources. Nowadays, research is focused on energy storage systems, such as batteries, hydrogen, and thermal storage, to reduce the curtailment of renewable energy sources and achieve flexible energy supply [4]. This hybrid renewable energy system (HRES), comprising both energy generation and storage components, serves as a feasible approach for clean energy shifting [5], for example, zero-energy/carbon buildings and communities in the architecture, engineering, and construction (AEC) field [6].

Although HRES is rapidly growing, its application still faces challenges: (1) the suitable sizing of HRES is a complex design process. It would be more complicated when facing diverse constraints and conflicting objectives. For instance, over- and undersizing of HRES could lead to redundant investment costs and less reliable configuration [7]. (2) As renewable energy penetration increases, the HRES system should be capable of reliability and sustainability. Thereby, long-duration energy storage and flexible energy consumption need to be considered by taking advantage of reasonable proportioning and operation of the storage system [8].

To conclude, sizing and operational optimization are critical for efficient and cost-effective HRES utilization. The purpose of this research is to develop an optimization workflow for HRES to improve its technological and economic

K. Qu · H. Zhang (✉)
School of Architecture, Southeast University, Nanjing, China
e-mail: 101002356@seu.edu.cn

Z. Peng (ed.), *Trends and Technological Challenges in Green Energy*, Green Energy and Technology, https://doi.org/10.1007/978-3-031-55579-4_9

performances. The sizing problem mainly focuses on the capacities of HRES, which comprises the PV-WT generation systems and a combined storage system with batteries for intra-day storage and hydrogen energy for seasonal storage. The three energy management strategies, namely maximizing self-consumption (MSC), time-of-use (TOU), and long-duration operational (LDO), are investigated and compared to explore their potential for performance improvements. In order to ensure the reliability and robustness of the proposed optimization framework, a single-family house is examined in both temperate (Nanjing) and tropical (Shenzhen) climate areas.

2 Methodology

2.1 System Structure and Sizing Variables

The structure of HRES is presented in Fig. 1. In this study, the on-site electricity generation by photovoltaics (PVs) and wind turbines (WTs) in direct current busbar could be converted to alternating current busbar to supply the load demand. The surplus power could be transformed into battery and hydrogen storage systems. When there is an energy gap, electricity will be delivered from the utility grid to

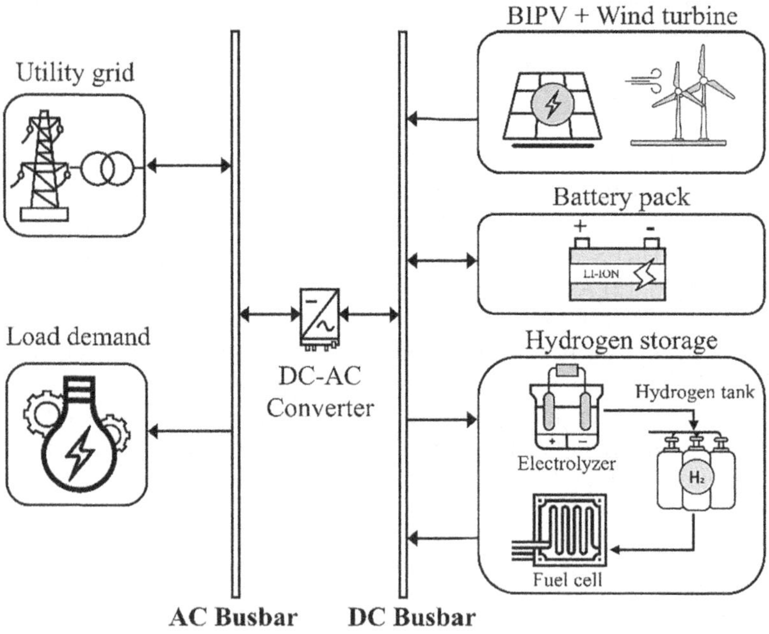

Fig. 1 Structure of the proposed hybrid renewable energy system

supply the insufficient energy. Moreover, the grid also serves as a backup to receive surplus electricity from the PV and WT, and the feed-in tariffs are also considered in this study.

2.2 System Component and Variables

Photovoltaic The electricity generation by PV panels is mainly affected by environmental temperatures and installation positions. The actual power generation, operating current, voltage, and temperature correction factor are calculated by Eqs. (1, 2, 3, and 4), respectively.

$$P_{\mathrm{PV}}^{t} = I^{t} V^{t} A_{\mathrm{PV}} \eta_{\mathrm{PV}}^{t} \eta_{\mathrm{s}} \eta_{\mathrm{PV}} / 1000 \tag{1}$$

$$I^{t} = I_{\mathrm{sc},0}\left(Rad^{t}/Rad^{0} - 1\right) + I_{\mathrm{pm},0} \tag{2}$$

$$V^{t} = V_{\mathrm{pm},0}\left(1 + 0.0593 lg\left(Rad^{t}/Rad^{0}\right)\right) \tag{3}$$

$$\eta_{\mathrm{PV}}^{t} = 1 - 0.005\left(T_{\mathrm{env}}^{t} + 30 Rad^{t}/800 - 25\right) \tag{4}$$

where A_{PV} is the area of PV panels. η_{s} is the performance degradations due to the dust and η_{pv} is the PV conversion efficiency. $I_{\mathrm{sc},\,0}$, $I_{\mathrm{pm},\,0}$ are the short circuit and peak currents of PV panels, respectively. $V_{\mathrm{pm},\,0}$ is the peak voltage. Rad^{t} and T_{env}^{t} are the solar radiation density and the outdoor environment temperature at time t.

Wind Turbine Instantaneous WT output power with respect to wind speed at any time is modeled as follows:

$$P_{\mathrm{WT}}^{t} = \begin{cases} 0, & v(t) < v_{\mathrm{ci}} < 0 \\ P_{\mathrm{WT,cap}} \cdot \dfrac{v^{t} - v_{\mathrm{ci}}}{v_{\mathrm{r}} - v_{\mathrm{ci}}}, & v_{\mathrm{ci}} < v^{t} < v_{\mathrm{r}} \\ P_{\mathrm{WT,cap}}, & v_{\mathrm{r}} < v^{t} < v_{\mathrm{co}} \end{cases} \tag{5}$$

$$v^{t} = v_{\mathrm{ref}}^{t}\left(\frac{Z}{Z_{\mathrm{ref}}}\right)^{\alpha} \tag{6}$$

where v^{t} is the environment wind speed, and v_{r}, v_{ci}, and v_{co} are the rated, cut-in, and cut-out wind speeds of WT, respectively. α is the wind shear coefficient. v_{ref}^{t} is the wind speed in meteorological data at time t, and Z and Z_{ref} are the WT hub height and wind speed measurement altitude, which are 6 and 10 m.

Battery Storage The operation of battery is represented in Eqs. (7 and 8).

$$E_{\mathrm{B}}^{t}=E_{\mathrm{B}}^{t-1}+P_{\mathrm{B,ch}}^{t}\eta_{\mathrm{ch}}-\frac{P_{\mathrm{B,dch}}^{t}}{\eta_{\mathrm{dis}}} \tag{7}$$

$$\begin{cases} E_{\mathrm{b,\,min}}\leq E_{\mathrm{B}}^{t}\leq E_{\mathrm{b,\,max}} \\ 0\leq P_{\mathrm{B,ch}}^{t}\leq P_{\mathrm{B,chmax}} \\ 0\leq P_{\mathrm{B,dch}}^{t}\leq P_{\mathrm{B,dchmax}} \end{cases} \tag{8}$$

where $P_{\mathrm{B,ch}}^{t}$ and $P_{\mathrm{B,dch}}^{t}$ are charge/discharge powers of battery at time t. η_{ch} and η_{dis} are their efficiencies. $E_{\mathrm{b,\,max}}$ and $E_{\mathrm{b,\,min}}$ are the upper and lower boundaries of the battery capacity. $P_{\mathrm{B,\,chmax}}$ and $P_{\mathrm{B,\,dchmax}}$ are power charge/discharge limits with 1 h.

Hydrogen Storage The operation of hydrogen storage is similar to battery storage as presented in Eqs. (9 and 10). However, their main difference is manifested in charging and discharging constraints, where the power transformed (P_{EL}^{t}) to and consumed (P_{FC}^{t}) from the hydrogen tank cannot exceed the rated power of the electrolyzer ($P_{\mathrm{EL,\,cap}}$) and fuel cell ($P_{\mathrm{FC,\,cap}}$).

$$E_{\mathrm{HT}}^{t}=E_{\mathrm{HT}}^{t-1}+P_{\mathrm{EL}}^{t}\eta_{\mathrm{EL}}-\frac{P_{\mathrm{FC}}^{t}}{\eta_{\mathrm{FC}}} \tag{9}$$

$$\begin{cases} E_{\mathrm{HT,\,min}}\leq E_{\mathrm{HT}}^{t}\leq E_{\mathrm{HT,\,max}} \\ 0\leq P_{\mathrm{EL}}^{t}\leq P_{\mathrm{EL,cap}} \\ 0\leq P_{\mathrm{FC}}^{t}\leq P_{\mathrm{FC,cap}} \end{cases} \tag{10}$$

where the η_{EL} and η_{FC} represent the efficiency of electrolyzer and fuel cell.

In this study, the sizing variables mainly include the components' capacity in HRES. However, it should be noted that PV panels are installed on the building envelope to form building integrated photovoltaics (BIPV), with a maximum capacity of 36 kWp. Therefore, the PV capacity is converted to the PV installation ratios on the rooftop and west and east facades (discussed in Sect. 3.3). Moreover, the battery and hydrogen tank (HT) capacities are converted to the total storage capacity and allocation proportion for holistically exploring the optimal sizing and allocation of storage components in HRES.

In summary, seven sizing variables are defined and optimized in this study. They are PV installation ratios on rooftop ($r_{\mathrm{PV,\,R}}$) and east–west facades ($r_{\mathrm{PV,\,WE}}$), the capacities of WT ($P_{\mathrm{WT,\,cap}}$), electrolyzer (EL, $P_{\mathrm{EL,\,cap}}$), and fuel cells (FC, $P_{\mathrm{FC,\,cap}}$), as well as the total storage capacity (S_{storage}) and allocation proportion (p_{storage}).

2.3 Energy Management Strategies and Variables

Maximizing Self-Consumption (MSC) Strategy As shown in Fig. 2a, MSC would transfer the excessive renewable power into the energy storage system (battery and hydrogen in this study) given that there is sufficient storage capacity [9]. Correspondingly, once the renewable energy fails to meet the load demand, the storage system will discharge to fill the energy gap. Renewable power beyond the maximal capacities of energy storage systems will be fed into the power grid for economic profits. In this way, the utilization of the energy storage system will be maximized.

Time-of-Use (TOU) Strategy The TOU is designed to maximize economic benefits by taking advantage of storing low-cost power during valley periods and releasing it during off-valley periods (see Fig. 2b) [10]. Specifically, as for the charging condition, if time t falls in the valley period, the excessive renewable energy will be supplied to hydrogen storage and then the battery. Moreover, if renewable energy fails to reach the storage limit of the battery, additional power will be purchased from the utility grid to charge the battery. By contrast, when the discharging time t falls in the valley period, both battery and hydrogen storage would not operate, and the load demand is only met by grid electricity.

Long-Duration Operational (LDO) Strategy The ODL is developed based on the MSC and TOU strategies and is examined to explore the potential of hydrogen for long-term storage. As shown in Fig. 2c, another four design conditions are defined in optimization process with ODL strategy. The t_{start} and t_{end} are time variables indicating the start and end time of the period with abundant renewable energy sources, and the $Limit_{abu}$ and $Limit_{scarce}$ are discharging limits of the hydrogen storage system in corresponding months. FC only operates to supply electricity when the stored energy in the hydrogen tank reaches its limits. It should be noted that these four variables in ODL can be determined by optimization algorithms. Thereby, the LDO strategy could dispatch the hydrogen energy storage considering seasonal power mismatch and hence improve the RES's technological efficiency over a yearly duration.

3 Optimization Method

3.1 Objective Function

Self-Sufficiency Ratio (SSR) The SSR indicates the percentage of load supplementation by on-site RES relative to total demand loads [11]. An SSR of 100% means that the HESS could realize the targets of seasonal storage and off-grid operation. Moreover, since the case study is in China, where the electricity from

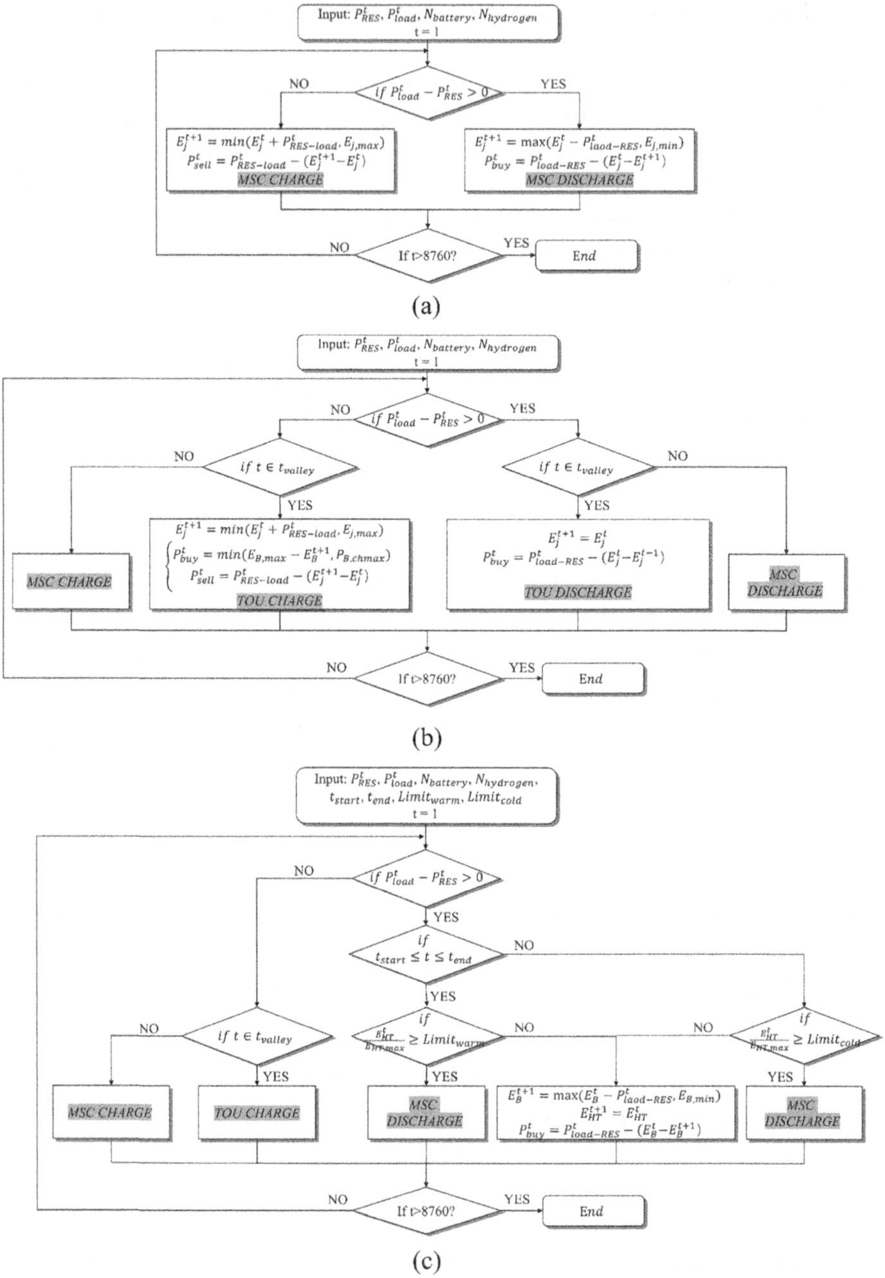

Fig. 2 Flowcharts for (**a**) MSC, (**b**) TOU, and (**c**) LDO strategies over a year

the utility grid is carbon intensive, the higher SSR also shows smaller environmental impacts. The calculation of SSR is expressed as

$$\mathrm{SSR} = \left(\sum\nolimits_{t=1}^{8760} P^t_{\mathrm{PV2G}} + P^t_{\mathrm{WT2G}} + P^t_{\mathrm{H}} + P^t_{\mathrm{B}} / \sum\nolimits_{t=1}^{8760} P^t_{\mathrm{load}}\right) \times 100\% \tag{11}$$

where P^t_{PV}, P^t_{WT}, P^t_{H}, and P^t_{B} are the power generated by PV, WT, hydrogen storage system, and battery to meet the load demand, respectively. P^t_{load} is the annual total load demand.

Total Annual Net Present Cost (TANPC) The TANPC indicates the annual cost of HESS considering its lifecycle and time value of money [12]. It concludes three ANPCs of investment, replacement, and salvage, as well as the operation and maintenance (O&M) cost. Furthermore, the annual electricity bill (AEB) consisting of electricity purchase and feed-in-tariff profit is also considered. Their calculations are expressed as

$$\mathrm{TANPC} = \frac{1}{\mathrm{PVF}} \sum\nolimits_{j=1}^{7} N^j \left(C^j_{\mathrm{I}} + \mathrm{PVF}^j_{\mathrm{R}} C^j_{\mathrm{R}} + \mathrm{PVF}^j_{\mathrm{S}} C^j_{\mathrm{S}}\right) + \mathrm{O\&MC} + \mathrm{AEB} \tag{12}$$

$$\mathrm{PVF} = (1+i)^{\mathrm{LF}} - 1/i(1+i)^{\mathrm{LF}} \tag{13}$$

$$\mathrm{PVF_R} = \begin{cases} 0, & \mathrm{LF}^j \geq \mathrm{LF} \\ (1+i)^{\left(\mathrm{LF}-\mathrm{LF}^j\right)} - 1/\left((1+i)^{\mathrm{LF}} - 1\right)(1+i)^{\left(\mathrm{LF}-\mathrm{LF}^j\right)}, & \mathrm{LF}^j < \mathrm{LF} \end{cases} \tag{14}$$

$$\mathrm{PVF_S} = (1+i)^{\mathrm{LF}} - 1/\left((1+i)^{\left(\mathrm{LF}-\mathrm{LF}^j\right)} - 1\right)(1+i)^{\mathrm{LF}} \tag{15}$$

$$\mathrm{O\&MC} = \sum\nolimits_{j=1}^{7} N^j C^j_{\mathrm{OM}} \tag{16}$$

where N^j and C^j are the capacities and unit costs of HESS's equipment j, respectively. PVF, $\mathrm{PVF_R}$, and $\mathrm{PVF_S}$ represent present value factors and are used to convert the initial, replacement, and salvage costs to net present costs considering equipment's lifetimes. LF and LF^j are years of the analyzing and equipment's lifetime, and i is the interest rate of 6% in this study.

$$\mathrm{AEB} = \sum\nolimits_{t=1}^{8760} \left(C^t_{\mathrm{im}} E^t_{\mathrm{im}} - C^t_{\mathrm{fit}} E_{\mathrm{ex}}\right) \tag{17}$$

where C^t_{im} and C^t_{fit} represent the unit prices of electricity purchase at hour t and feed-in-tariff, respectively. E^t_{im} and E_{ex} represent the purchase and sold electricity from the grid at the k_{th} hour.

3.2 Optimization Algorithm

This study adopts the nondominated sorting genetic algorithm II (NSGA-II) as the optimization solution. This algorithm is a modified genetic algorithm and has been proven robust and effective in tackling multiobjective problems by exploring nondominated (Pareto) solutions, which are feasible solutions that cannot improve one objective without deteriorating other objectives. Its core mechanism lies in crowding distance and nondominated sorting, which ensures diversity and uniform distribution for generating populations (design configurations). Table 1 shows the parameter settings of NSGA-II, and the optimization process was implemented using Wallacei software (https://www.wallacei.com/). Interested readers can refer to [13, 14] for more details about the algorithm and its applications.

3.3 Specifications of Case Study

The case study is a grid-connected single-family house located in Nanjing, China, a hot-summer and cold-winter zone. The building is equipped with a 10-kWh lithium battery and a total of 36 kWp of PV panels installed on its roof and west and east façades. An air-source heat pump is used for cooling and heating. Since this house is not put into use, its operational parameters and schedules are set according to Chinese general code GB 55015-2021 [15], and the annual building load is simulated by EnergyPlus as 7259.18 kWh. The operation of the household RES adopts the MSC. As a baseline, the existing case house has a self-sufficient ratio of 66.5% with a TANPC of 543.36 USD. The electricity supplied by the utility grid are 0.05, 0.11 and 0.18 USD in valley (0:00–5:00, 21:00–24:00), flat (6:00–7:00, 11:00–17:00), and peak (8:00–10:00, 18:00–20:00) periods, respectively. The basic information of the HRES components is listed in Table 2.

Table 1 Parameter settings of NSGA-II

	Population	Iteration	Crossover	Mutation	Crossover distribution	Mutation distribution
Value	100	50	0.9	1/variable number	20	20

Table 2 Case study parameters of HRES components

	PV	WT	Battery	EL	HT	FC	Converter
Lifetime (years)	20	20	5	15	20	5	10
Investment (USD/kW)	500	571	260	200	62	700	300
Replacement factor (%)	90	90	90	90	90	90	90
Salvage factor (%)	12	2	8	8	20	5	2
O&M factor (%)	2	2	2	1.5	0.5	5	0.5

Table 3 Optimization parameters and baseline for two cases

Cases	Objective	Strategy	Variable
1- N, S	SSR, TANPC	MSC	$r_{PV,\ R}$, $r_{PV,\ WE}$, $P_{WT,\ cap}$, $P_{EL,\ cap}$, $P_{FC,\ cap}$, $S_{storage}$, $p_{storage}$
2- N, S		TOU	$r_{PV,\ R}$, $r_{PV,\ WE}$, $P_{WT,\ cap}$, $P_{EL,\ cap}$, $P_{FC,\ cap}$, $S_{storage}$, $p_{storage}$
3- N, S		LDO	$r_{PV,\ R}$, $r_{PV,\ WE}$, $P_{WT,\ cap}$, $P_{EL,\ cap}$, $P_{FC,\ cap}$, $S_{storage}$, $p_{storage}$, t_{start}, t_{end}, $Limit_{warm}$, $Limit_{cold}$
Baseline-N	66.5%, 543 USD	MSC	$r_{PV,\ R}$ (100%), $r_{PV,\ WE}$ (100%), $P_{WT,\ cap}$ (0 kW), $P_{EL,\ cap}$ (0 kW), $P_{FC,\ cap}$ (0 kW), $S_{storage}$ (10 kWh), $p_{storage}$ (100%)
Baseline-S	68.64%, 574 USD	MSC	

N and S represent Nanjing and Shenzhen, respectively

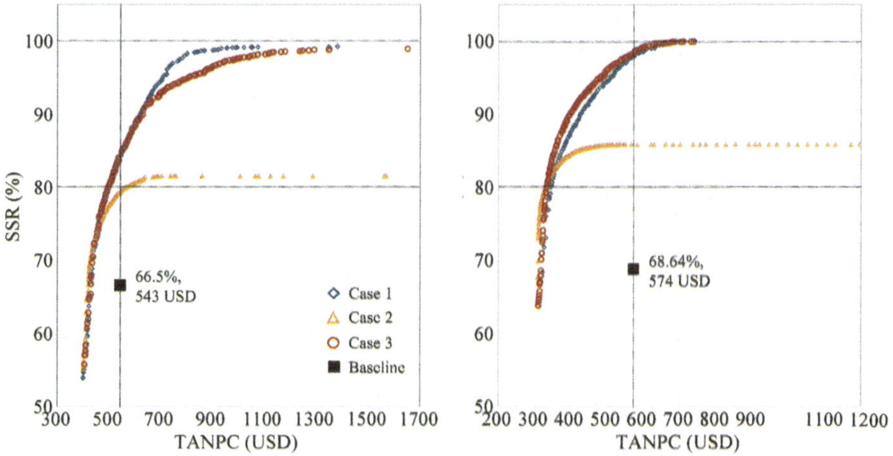

Fig. 3 SSR and TANPC of Pareto solutions in Nanjing (left) and Shenzhen (right)

To further demonstrate the proposed optimization method, case building is also investigated in Shenzhen (a hot summer and warm winter area) as a hypothetical scenario. In summary, a total of six simulation cases are examined in this study. Their parameter settings and baseline data are listed in Table 3.

4 Results and Discussion

4.1 Analysis of the Pareto Frontiers

Figure 3 shows the Pareto fronts of cases in both Nanjing and Shenzhen locations. It can be seen that all Pareto solutions present better performance than baselines for both technological and economic indicators. Specifically, based on the equivalent baseline TANPCs, the maximal SSRs of optimized configurations could reach 84.19% in Nanjing and 98.49% in Shenzhen, which outperform baselines by 17.69% and 29.85%, respectively.

Moreover, the performance of the TOU strategy in optimization case 2 is the lowest, especially in exploring SSR. Specifically, the maximal SSRs for case 2 in Nanjing and Shenzhen are 81.66% and 85.99%, whereas cases 1 and 3 could reach almost 100% SSR.

Comparison between cases 1 and 3 reveals that case 1 (MSC strategy) is better suited to exploring the Pareto front in Nanjing (see Fig. 3, left). In particular, case 1 could achieve higher renewable energy penetration with equivalent TANPC in the SSR range from 90% to 99%. Comparatively, the proposed LDO strategy would be more appropriate for Shenzhen, where case 3 would save 30 USD compared to case 1 at around 90% SSR. As a result, location and climate conditions can significantly influence HRES operations. The proposed LDO strategy is more suitable for tropical climate zones with high-energy mismatches than for temperate zones.

4.2 System Configuration Comparison

Table 4 presents the selected HRES configurations and their parameters. The configurations are chosen with similar SSRs of 80% and 100%. As for the power generation components, it can be seen that all selection configurations maximize the PV installation ratio on the rooftop and the WT capacity. Accordingly, complementary power generation may compensate for the intermittent nature of RES and result in lower variability and higher HRES utilization rates as compared to single PV or WT generation profiles.

As for the storage systems, when the seasonal storage is required (SSR equal to nearly 100%), the total storage capacities increase from around 30 to 99 kWh in Nanjing and from around 10 to 60 kWh in Shenzhen. Moreover, the share of hydrogen energy in the overall storage system would also increase. For instance, the hydrogen proportion of case 3 in Shenzhen with 80% SSR is 7% and increases to 77% as SSR equals to 100%. Hydrogen is therefore considered to be an essential long-term energy storage option if seasonal storage and off-grid operation are required.

In addition, a twofold increase in TANPCs is observed in cases 1 and 3 when SSR is increased from 80% to 100%. As marginal investments increase, HRES is facing challenges in improving its technological efficiency. Facing this problem, the proposed optimization method could be useful since it can generate a series of design configurations under different cost scenarios for decision-makers to choose according to their budget.

Table 4 The variables and objectives of selected solutions. The 80% and maximum SSRs are set as the selection criteria

	Nanjing					Shenzhen				
Case	1-MSC	2-TOU	3-LDO	1-MSC	3-LDO	1-MSC	2-TOU	3-LDO	1-MSC	3-LDO
$r_{PV,\ R}$ (%)	100 (16.63kWp)									
$r_{PV,\ WE}$ (%)	31 (6 kWp)	67 (13.98 kWp)	0	98 (18.98 kWp)	93 (18 kWp)	27 (5.23 kWp)	1 (0.2 kWp)	1 (0.2 kWp)	99 (19.17 kWp)	78 (15.11 kWp)
$P_{WT,\ cap}$ (kW)	9.8	9.92	9.96	9.7	9.93	9.97	9.99	9.97	9.99	9.99
$P_{EL,\ cap}$ (kW)	0.45	0.09	0.49	3.32	6.2	0.01	0.01	0.14	2.87	2.88
$P_{FC,\ cap}$ (kW)	0.19	0.59	0.3	4.12	4.91	0.01	0.01	0.02	5.3	5.64
$S_{storage}$(kWh)	30.58	29.31	33.69	**99.2**	**98.99**	7.55	10.79	6.25	**54.9**	**67.96**
HT (kWh)	10.7 (65%)	12.6 (43%)	23.58 (70%)	**72.42 (73%)**	**83.15 (84%)**	1.51 (20%)	3.02 (28%)	0.44 (7%)	**39.53 (72%)**	**52.33 (77%)**
Battery (kWh)	19.88 (35%)	16.71 (57%)	10.11 (30%)	26.78 (27%)	15.84 (16%)	6.04 (80%)	7.77 (72%)	5.81 (93%)	15.37 (28%)	15.63 (23%)
t_{start}	–	–	**244**	–	**4238**	–	–	**4427**	–	**5421**
t_{end}	–	–	**752**	–	**7468**	–	–	**4322**	–	**4646**
$Limit_{abu}$(%)	–	–	9	–	1	–	–	94	–	23
$Limit_{scarce}$(%)	–	–	0	–	16	–	–	74	–	5
TANPC (USD)	501	562	498	932	1350	344	328	333	738	737
SSR (%)	80.1	80.04	80.22	99.04	98.92	80.1	80	80.05	100	100

5 Conclusions

In this study, an optimization study is presented for sizing and operational problems of hybrid renewable energy systems consisting of PV/WT generation and hydrogen/battery storage units, both in temperate and tropical climates. Some main conclusions include the following:

- As a result of the optimization method, HRES's efficiency can be improved from the perspective of both technology and economics. The optimized solutions could achieve 17.69% and 29.85% SSR improvements with the equivalent annual net present costs as the baselines.
- The proposed LDO strategy performs well in optimizing HRES; however, it may be impacted by location. Compared to temperate climate zones, it has better performance in tropical climate regions with higher seasonal energy mismatch.
- Results reveal that the need for seasonal storage leads to an increase in both the capacity and proportion of hydrogen storage. In the case of Nanjing, this increase amounts to 8% and 14% under the MSC and ODL strategies, respectively. In Shenzhen, the increase is even more significant, ranging from 52% to 70%. Given hydrogen's capacity for long-term energy storage, its role is considered crucial for both seasonal storage or off-grid operations.
- The marginal benefit of investment costs for HRES would diminish afterward, making it more than twofold cost to realize 100% renewable energy penetration from 80% SSR. Accordingly, the trade-off between tail investment and total renewable penetration should be balanced in terms of standards and practical requirements. An example of a reasonable investment would be near-zero-energy buildings with 90% SSR.

As the method and results of this study show, researchers and engineers may be able to develop hybrid renewable energy systems for low/net-zero carbon communities and microgrids.

References

1. M. D. A. Al-falahi, S. D. G. Jayasinghe, and H. Enshaei, "A review on recent size optimization methodologies for standalone solar and wind hybrid renewable energy system," Energy Convers. Manag., vol. 143, pp. 252–274, 2017, https://doi.org/10.1016/j.enconman.2017.04.019.
2. A. F. Lyden, C. S. Brown, I. Kolo, G. Falcone, and D. Friedrich, "Seasonal thermal energy storage in smart energy systems: District-level applications and modelling approaches," Renew. Sustain. Energy Rev., vol. submitted, no. June, 2022.
3. T. M. Alabi et al., "A review on the integrated optimization techniques and machine learning approaches for modeling, prediction, and decision making on integrated energy systems," Renew. Energy, vol. 194, pp. 822–849, 2022, https://doi.org/10.1016/j.renene.2022.05.123.

4. W. Hu, Q. Shang, X. Bian, and R. Zhu, "Energy management strategy of hybrid energy storage based on Pareto optimality," Appl. Energy, vol. 17, no. October, pp. 169–175, 2022, https://doi.org/10.1093/ijlct/ctab094.
5. V. Khare, S. Nema, and P. Baredar, "Solar-wind hybrid renewable energy system: A review," Renew. Sustain. Energy Rev., vol. 58, pp. 23–33, 2016, https://doi.org/10.1016/j.rser.2015.12.223.
6. J. Liu, X. Chen, H. Yang, and K. Shan, "Hybrid renewable energy applications in zero-energy buildings and communities integrating battery and hydrogen vehicle storage," Appl. Energy, vol. 290, no. March, p. 116733, 2021, https://doi.org/10.1016/j.apenergy.2021.116733.
7. N. Niveditha and M. M. Rajan Singaravel, "Optimal sizing of hybrid PV–Wind–Battery storage system for Net Zero Energy Buildings to reduce grid burden," Appl. Energy, vol. 324, no. April, p. 119713, 2022, https://doi.org/10.1016/j.apenergy.2022.119713.
8. T. S. Le, T. N. Nguyen, D. K. Bui, and T. D. Ngo, "Optimal sizing of renewable energy storage: A techno-economic analysis of hydrogen, battery and hybrid systems considering degradation and seasonal storage," Appl. Energy, vol. 336, no. July 2022, p. 120817, 2023, https://doi.org/10.1016/j.apenergy.2023.120817.
9. B. Zou et al., "Energy management of the grid-connected residential photovoltaic-battery system using model predictive control coupled with dynamic programming," Energy Build., vol. 279, p. 112712, 2023, https://doi.org/10.1016/j.enbuild.2022.112712.
10. D. Sadeghi et al., "Designing, optimizing and comparing distributed generation technologies as a substitute system for reducing life cycle costs, CO2 emissions, and power losses in residential buildings," Energy, vol. 253, p. 123947, 2022, https://doi.org/10.1016/j.energy.2022.123947.
11. Y. Zhang, T. Ma, and H. Yang, "Grid-connected photovoltaic battery systems: A comprehensive review and perspectives," Appl. Energy, vol. 328, no. September, p. 120182, 2022, https://doi.org/10.1016/j.apenergy.2022.120182.
12. A. Azad and H. Shateri, "Design and optimization of an entirely hybrid renewable energy system (WT/PV/BW/HS/TES/EVPL) to supply electrical and thermal loads with considering uncertainties in generation and consumption," Appl. Energy, vol. 336, no. December 2022, p. 120782, 2023, https://doi.org/10.1016/j.apenergy.2023.120782.
13. K. Deb, A. Pratap, S. Agarwal, and T. Meyarivan, "A fast and elitist multiobjective genetic algorithm: NSGA-II," IEEE Trans. Evol. Comput., vol. 6, no. 2, pp. 182–197, 2002, https://doi.org/10.1109/4235.996017.
14. M. Hamdy, A. T. Nguyen, and J. L. M. Hensen, "A performance comparison of multi-objective optimization algorithms for solving nearly-zero-energy-building design problems," Energy Build., vol. 121, pp. 57–71, 2016, https://doi.org/10.1016/j.enbuild.2016.03.035.
15. Ministry of Housing and Urban-Rural Development of the People's Republic of China. General code for energy efficiency and renewable energy application in buildings (GB55015-2021), 2021 (in Chinese).

Numerical Simulation About Transient Performance of Regenerative Cooling System for Combustion Chamber

Chen Zhang, Hui Gao, Jiajun Zhao, and Dongsheng Wen

1 Introduction

Hypersonic technologies have become one of the most challenging fields in recent years. One of the most prominent problems of hypersonic vehicles (Ma > 5.0) is the thermal risk [1, 2]. The huge thermal loads imposed by the aerodynamic heating and combustion put high demands on the performance of heat-resistant materials and cooling systems. The cooling systems can be classified into two types: passive cooling and active cooling [3]. The passive cooling system mainly refers to the heat insulation materials and structure, which are heavy and expensive. The active cooling methods include film cooling, transpiration cooling, regenerative cooling, and their combinations. Regenerative cooling uses onboard hydrocarbon fuels as

C. Zhang
School of Aeronautic Science and Engineering, Beihang University, Beijing, People's Republic of China

Zhongfa Aviation Institute, Beihang University, Hangzhou, China

H. Gao (✉)
School of Aeronautic Science and Engineering, Beihang University, Beijing, People's Republic of China

Zhongfa Aviation Institute, Beihang University, Hangzhou, China

Ningbo Institute of Technology, Beihang University, Ningbo, People's Republic of China
e-mail: h.gao@buaa.edu.cn

J. Zhao
School of Aeronautic Science and Engineering, Beihang University, Beijing, People's Republic of China

Ningbo Institute of Technology, Beihang University, Ningbo, People's Republic of China

D. Wen
Institute of Thermodynamics, Technical University of Munich, Munich, Germany

Z. Peng (ed.), *Trends and Technological Challenges in Green Energy*, Green Energy and Technology, https://doi.org/10.1007/978-3-031-55579-4_10

coolant, which is neat and efficient. The endothermic hydrocarbon fuels (EHFs) first flow through the wall of the hot parts, absorbing heat and decomposing into small species. Then, it is injected into the combustion chamber. As the heat absorbed by the endothermic hydrocarbon fuels is not wasted, it forms the concept of "regenerative." During the cooling process, the EHF often works in supercritical state to avoid phase change and boiling heat transfer. In addition, the pyrolysis reactions further change the composition of the fluid, making the flow and heat transfer process extremely complex [4]. CFD simulation as an efficient method for predictive study of complex fluid system has become a widely adopted method for cooling performance estimation [5], structural design [6], and coking performance evaluation [7].

Most of the former numerical research of regenerative cooling system are based on steady-state simulation with evenly distributed thermal boundary, which is different from the real case. For regenerative cooling in the combustion chamber, the thermal boundary is determined by the state of combustion. Currently, most of the hypersonic vehicles are equipped with scramjet engine. To date, one of its distinguishing features for scramjet-driven hypersonic vehicles is that the scramjet cannot be initiated at low speed. Consequently, the hypersonic vehicles are often carried to a certain altitude first and then activated [8]. Considering the real process, it is necessary to estimate the transient performance of regenerative cooling system before/after ignition of scramjet engine. Bao et al. [9] set an unevenly distributed thermal boundary to study the effect of working conditions on the heat sink of endothermic hydrocarbon fuels. But the transient process was not revealed. The real thermal boundary imposed by supersonic combustion can be obtained via RANS or LES simulation [10]. However, the inherent complexity of the combustion process leaves a distance between numerical simulation and reality.

In this study, the transient characteristics of regenerative cooling system using EHF are numerically simulated using the ANSYS Fluent software. The start-up performance of regenerative cooling system is simulated. Before ignition, the thermal load is induced by aerodynamic heating. Then, the thermal boundary is replaced by unevenly distributed heat flux. To evaluate the effect of real thermal boundaries, the flow and heat transfer characteristics are compared with equivalent uniform heat flux. With the relaxation time of regenerative cooling system after ignition, transient distribution of wall temperature is obtained. The temperature and conversion of EHF at the outlet at different times are calculated.

2 Numerical Methods

In this study, the CFD simulation is performed based on the commercial software ANSYS Fluent, which is a suitable tool for the numerical simulation of complex fluid system including chemical reactions.

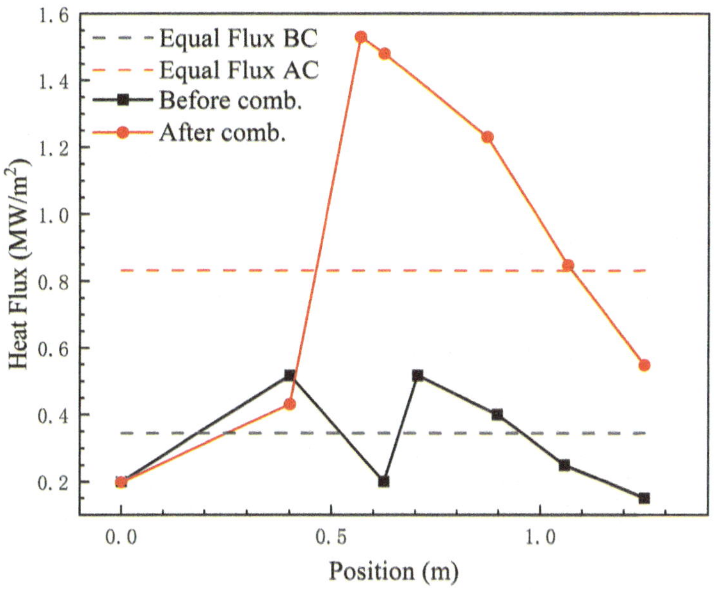

Fig. 1 Detailed profiles of heat flux boundary

2.1 *Geometry and Boundary Conditions*

A circular tube is selected for the cooling performance estimation. The length of the tube is 1250 mm with a diameter of 2 mm. The circular tube is simulated by a two-dimensional axisymmetric model. The EHF is fed with a constant mass flow rate, 2.0 g s^{-1}. The fuel flows into the tube at a temperature of 373.15 K. The operating pressure is 3.45 MPa to ensure the supercritical state of EHF. In this study, three thermal boundaries are applied, that is, uniform heat flux 0.35 MW m^{-2} for aerodynamic heating, distributive heat flux for combustion, and another uniform one for equivalent heat flux, which is around 0.85 MW m^{-2}. The detailed profiles of the thermal boundary are shown in Fig. 1 according to a ground test of scramjet engine [11]. Before combustion, the thermal load is imposed by the aerodynamic heating; see the relatively stable heat flux in Fig. 1. After combustion, the heat flux rises rapidly after the fuel injection nozzle, see the red line in Fig. 1.

2.2 *Governing Equations*

The transient governing equations are listed in this section. For steady-state simulation, the transient terms are dismissed, and the governing equations become steady state.

For mass conservation:

$$\frac{\partial \rho}{\partial t} + \nabla \cdot \left(\rho \vec{u}\right) = 0 \tag{1}$$

For conservation of momentum:

$$\frac{\partial}{\partial t}\left(\rho \vec{u}\right) + \nabla \cdot \left(\rho \vec{u}\,\vec{u}\right) = -\nabla p + \nabla \cdot \overline{\overline{\tau}}_{\text{eff}} \tag{2}$$

For conservation of energy:

$$\frac{\partial}{\partial t}(\rho e_t) + \nabla \cdot \left(\rho \vec{u}\, e_t + p\vec{u}\right) = \nabla \cdot (\lambda_{\text{eff}} \nabla T) \tag{3}$$

The pyrolysis of EHF changes the components of the fluid mixture; to describe the multispecies transport phenomenon, the species transport equation is solved:

$$\frac{\partial}{\partial t}(\rho Y_i) + \nabla \cdot \left(\rho Y_i \vec{u}\right) = -\nabla \cdot \left(\rho Y_i \vec{u}_{d,i}\right) + S_i \tag{4}$$

The source term S_i is calculated by the following equation:

$$S_i = \omega_i M_{\omega i} \tag{5}$$

The SST k-ω model is selected to describe the flow and heat transfer in turbulent regime:

$$\frac{\partial}{\partial t}(\rho k) + \nabla \cdot \left(\rho \vec{u}\, k\right) = \nabla \cdot \left[\left(\mu + \frac{\mu_t}{\sigma_k}\right)\nabla k\right] + G_k - Y_k \tag{6}$$

$$\frac{\partial}{\partial t}(\rho \omega) + \nabla \cdot \left(\rho \vec{u}\, \omega\right) = \nabla \cdot \left[\left(\mu + \frac{\mu_t}{\sigma_\omega}\right)\nabla \omega\right] + G_\omega - Y_\omega + D_\omega \tag{7}$$

2.3 *Chemical Reactions and Thermophysical Properties*

In this study, the EHF is represented by a pure material, that is, n-decane, as it possesses similar physical properties to the real EHF. To describe the pyrolysis behavior, a widely accepted global model, PPD model, is adopted [12]. Eighteen pyrolytic products are generated, and its distribution is considered constant under different operating conditions. The kinetics of the pyrolysis process is described by Arrhenius form. Detailed information on the reaction parameters can be found in the reference, which is not reproduced in this study. As mentioned before, n-decane works in supercritical pressure. During the cooling process, its thermophysical properties change drastically, which significantly affects the cooling performance. In this study, the SUPERTRAPP from NIST is selected to calculate the density, heat capacity, thermal conductivity, and viscosity of n-decane and its pyrolytic products.

3 Results and Discussion

3.1 Model Validation and Mesh Independence

The validation of the PPD model has been performed in our former research [13]. The mesh independence is tested in the first place. Three meshes with different axial divisions are compared, which is shown in Fig. 2. The other parameters of the meshes are controlled the same as the numerical error generated by the mesh is mainly from the transfer phenomena near the wall, that is, the boundary layer.

The detailed parameters of the meshes are listed in Table 1, including the sectional division, initial height, growth rate, and the number of longitude divisions. To ensure the correct simulation by using SST k-ω turbulence model, the y+ value near the wall is checked to be lower than 1.0. The temperature and n-decane

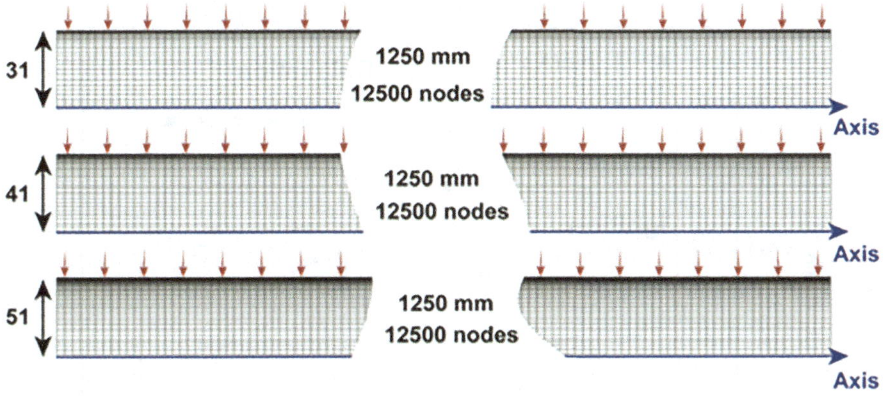

Fig. 2 Three meshes for mesh independence test

Table 1 Detailed information of three meshes

Mesh number	Sectional division in fluid domain	Initial height	Growth rate	Number of longitude division
Mesh 1	31	2 μm	1.05	12,500
Mesh 2	41	2 μm	1.05	12,500
Mesh 3	51	2 μm	1.05	12,500

conversion at the outlet are compared. Results show that the relative error is within 1%. Considering computational accuracy and efficiency, the No. 1 mesh is selected for the following simulation.

3.2 Transient Performance During Ignition Process

The temperature of bulk fluid is shown in Fig. 3. The ignition delay time is set to be 0.01 s according to Ruan et al. [14]. Before the ignition, the coolant is heated by the aerodynamic heat flux, and the bulk temperature of fluid at the whole fluid region is lower than the initial cracking temperature of n-decane. When the fuel is ignited and the heat flux gradually rises within the ignition delay time, the outlet temperature first decreases. This is attributed to the thermal decomposition of n-decane that is activated and the absorbed heat is larger than the increased heat. With the thermal load imposed by the combustion further rising, the temperature increases and becomes constant at around 0.175 s, which means the steady state is achieved.

The n-decane conversion of outlet is shown in Fig. 4. It is clear that when the fuel temperature exceeds the threshold value, the n-decane starts to decompose and the conversion rate grows dramatically and reaches 100% within 0.1 s.

The regenerative cooling system is mainly used to guarantee that the maximum temperature is lower than the limit of solid material. Therefore, the temperature of

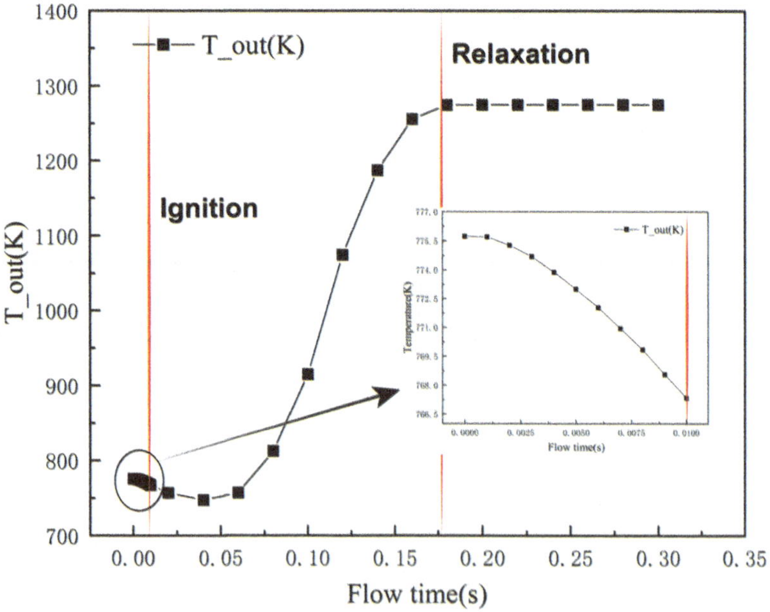

Fig. 3 Outlet temperature during the relaxation period

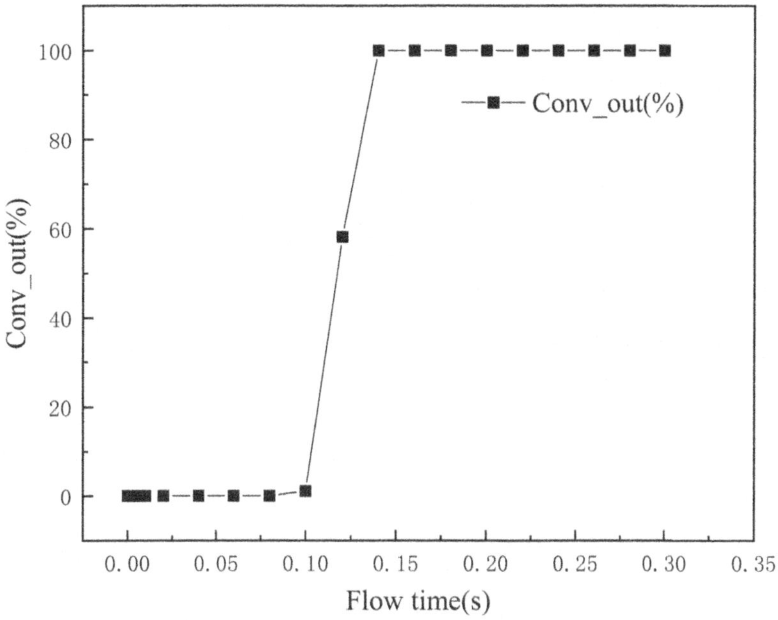

Fig. 4 Outlet temperature during the relaxation period

the heated wall within 10 ms is observed as depicted in Fig. 5. During the ignition process, the maximum temperature is increased by 150 K.

3.3 *Steady State Characteristics Comparison*

As mentioned before, the observed relaxation time is 0.175 s, and the cooling system becomes steady. To compare the real heat flux with evenly distributed one, the cooling performance under the equivalent heat flux, 0.85 MW m^{-2}, is simulated and compared. Figure 6 shows the contours of temperature and mass fraction of n-decane under aerodynamic heat flux, distributive heat flux imposed by combustion, and equivalent heat flux of real. Before the ignition, the n-decane is not sufficiently heated and the pyrolysis is not triggered by observing Fig. 6a, b. In Fig. 6a, the heat released by the combustion significantly leveled up the temperature after the ignition nozzle. As the maximum temperature is located near the peak of heat flux, the temperature field is quite different from that under the equivalent heat flux of combustion. As the higher temperature is conductive to the pyrolysis, there exists a local high conversion of n-decane at around 0.6 m as shown in Fig. 6b.

Similarly, the wall temperature profiles under three conditions are plotted in Fig. 7. Under real heat flux, the temperature distribution is different from the uniform

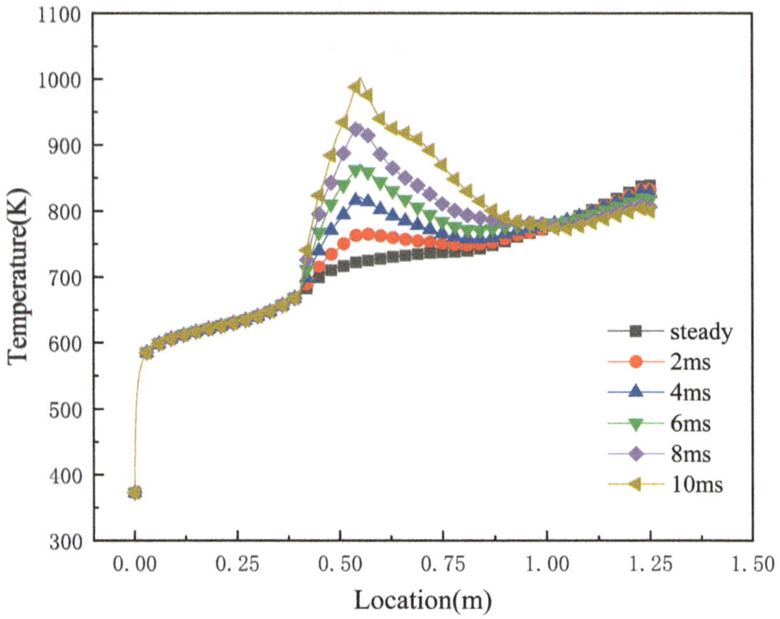

Fig. 5 Wall temperature distribution during the ignition process

thermal boundary, manifested as different maximum temperature and location of the peak temperature. Before the ignition nozzle, the thermal load is mainly generated by aerodynamic heating and temperature distribution is almost the same as the initial. In reality, the intensity of the heat transfer is determined mainly by temperature difference, and it should decrease with the increase in wall temperature (the decrease of temperature difference). However, under equivalent heat flux of combustion, the heat flux remains the same at any point on heat surface. Therefore, more heat is absorbed in the high-temperature region, leading to higher maximum temperature compared to that under distributive heat flux.

4 Conclusion

A CFD simulation about transient characteristics of regenerative cooling system using n-decane as coolant under real heat flux was done in this study. To reveal the influence of real flux on cooling performance, three thermal boundaries are applied for the numerical simulation and comparison. There are four main conclusions:

1. The real heat flux is characterized by aerodynamic heating and combustion. The profile of the real heat flux reflects both spatial and temporal distribution.

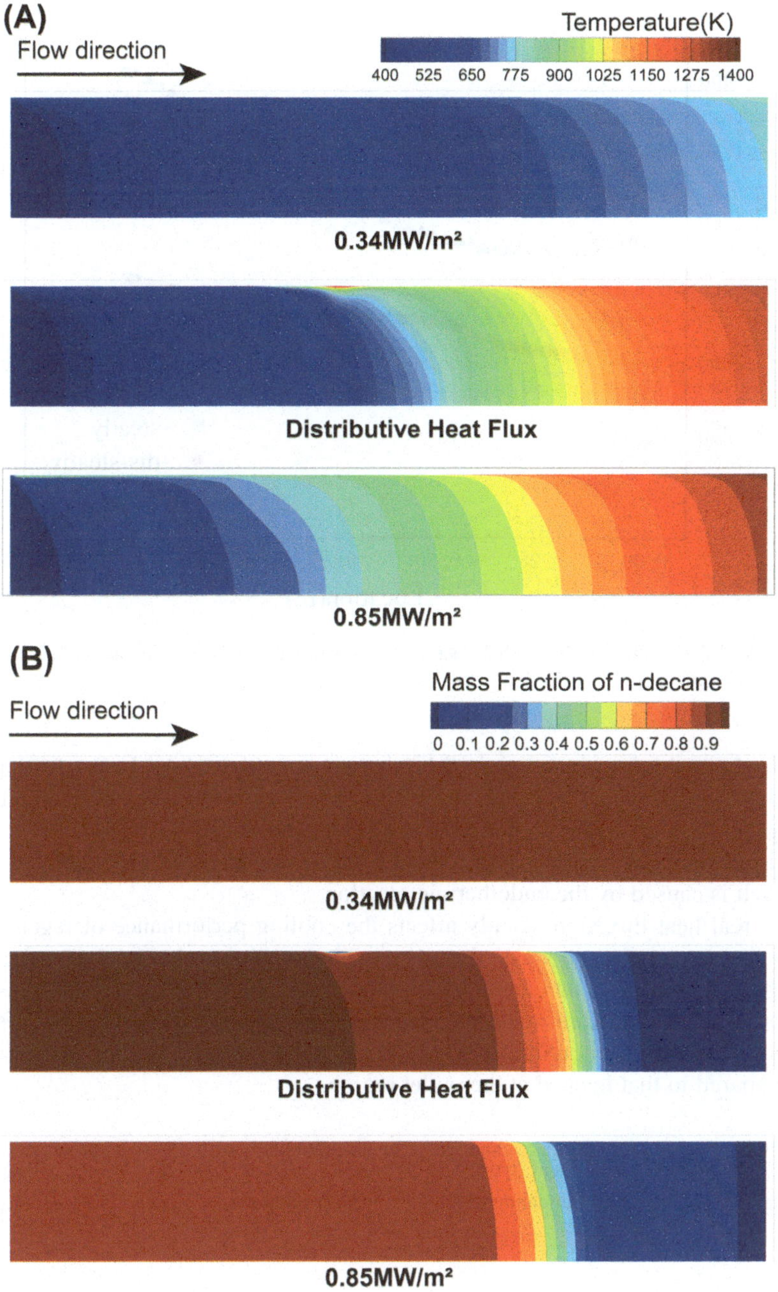

Fig. 6 Contours of (**a**) fluid temperature (**b**) mass fraction of n-decane before and after the combustion process

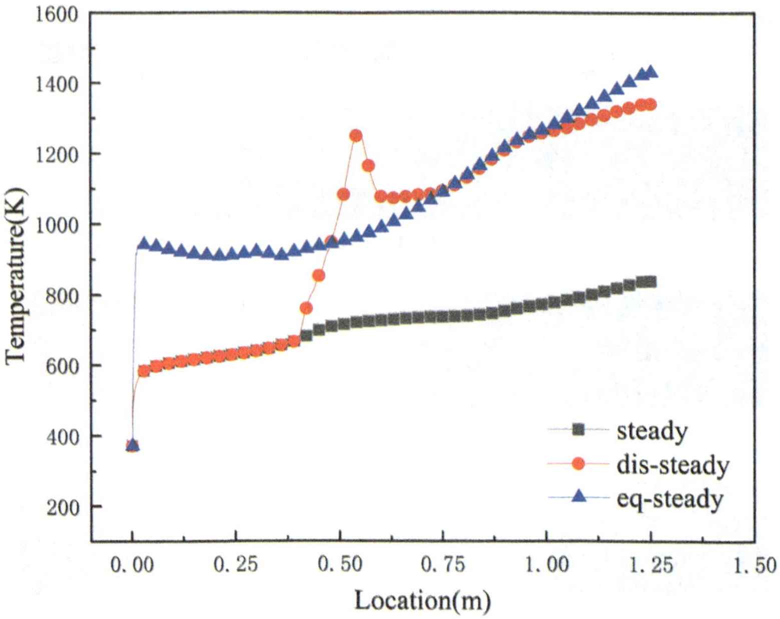

Fig. 7 Wall temperature profiles under aerodynamic heating, uniform combustion heat flux, and distributive combustion heat flux

2. The thermal relaxation time is 0.175 s in this study, which is longer than the ignition delay time, that is, 0.01 s. Therefore, the start-up characteristics of regenerative cooling system need to be considered.
3. During ignition, the temperature of the fuel first decreases and then increases, which is caused by the endothermic reaction.
4. The real heat flux significantly affects the cooling performance of regenerative cooling system in two ways: the chemical heat sink is enlarged due to the local high thermal flux; the temperature peaks at different locations, that is, the maximum temperature under distributive heat flux is almost 100 K lower than that under equivalent uniform heat flux, and it is located at the middle of the tube compared to that located at the outlet region.

Acknowledgments The authors acknowledge financial support from the Natural Science Foundation of Zhejiang Province (NO. KYZ1121004). The authors also acknowledge the support from the Academic Excellence Foundation of BUAA for PhD students.

References

1. E. T. Curran. Scramjet engines: the first forty years. Journal of Propulsion and Power 17(6): 1138–1148.(2001)
2. W. Ning, P. Yu,Z. Jin. Research status of active cooling of endothermic hydrocarbon fueled scramjet engine. Proceedings of the Institution of Mechanical Engineers, Part G: Journal of Aerospace Engineering 227(11): 1780–1794.(2013)
3. S. Luo, D. Xu, J. Song,J. Liu. A review of regenerative cooling technologies for scramjets. Applied Thermal Engineering 190: 116754.(2021)
4. F. Zhong, X. Fan, G. Yu, J. Li,C.-J. Sung. Heat transfer of aviation kerosene at supercritical conditions. Journal of Thermophysics and Heat Transfer 23(3): 543–550.(2009)
5. X. Li, S. Zhang, J. Qin,W. Bao. Parametric analysis on the thermal behavior of cracking hydrocarbon fuel flow inside asymmetry heated cooling channels with micro-ribs. International Journal of Heat and Mass Transfer 160: 120154.(2020)
6. Y. Feng, J. Cao, X. Li, S. Zhang, J. Qin,Y. Rao. Flow and Heat Transfer Characteristics of Supercritical Hydrocarbon Fuel in Mini Channels With Dimples. Journal of Heat Transfer 139(12).(2017)
7. C. Zhang, H. Gao, J. Zhao, J. Zhou, X. Ma,D. Wen. Dynamic coking simulation of supercritical n-decane in circular tubes. Fuel 331: 125859.(2023)
8. J. Hank, J. Murphy,R. Mutzman (2008). The X-51A scramjet engine flight demonstration program. 15th AIAA international space planes and hypersonic systems and technologies conference.
9. W. Bao, X. Li, J. Qin, W. Zhou,D. Yu. Efficient utilization of heat sink of hydrocarbon fuel for regeneratively cooled scramjet. Applied Thermal Engineering 33: 208–218.(2012)
10. F. Ladeinde. Advanced computational-fluid-dynamics techniques for scramjet combustion simulation. AIAA Journal 48(3): 513–514.(2010)
11. T. Hiraiwa, T. Kanda, K. Kobayashi,T. Saito. Experiments on a scramjet engine with ramp-compression inlet at Mach 8. Journal of Propulsion and Power 22(2): 440–446.(2006)
12. T. A. Ward, J. S. Ervin, S. Zabarnick,L. Shafer. Pressure Effects on Flowing Mildly-Cracked n-Decane. Journal of Propulsion and Power 21(2): 344–355.(2005)
13. C. Zhang, H. Gao, J. Zhao, J. Zhou,D. Wen. Cooling performance optimization of supercritical hydrocarbon fuel via bi-directional flow pattern. Applied Thermal Engineering 212: 118563. (2022)
14. B. Ruan, S. Huang, H. Meng,X. Gao. Flow dynamics in transient heat transfer of n-decane at supercritical pressure. International Journal of Heat and Mass Transfer 115: 206–215.(2017)

An Evolutionary Game Study on the Behavioral Strategies of Carbon Label Stakeholders Based on Prospect Theory

Chaoping Liu and Xiaomei Wu

1 Introduction

With the rapid growth of the global economy since the twenty-first century, the problem of climate change caused by the massive emission of greenhouse gases has become increasingly prominent. To cope with the climate crisis, countries around the world are actively looking for various initiatives, and the concept of low-carbon development has been birthed and gradually become a key strategy initiative [1]. Among them, the carbon labeling scheme has been promoted and applied in many countries for its effectiveness in energy-saving, emission reduction, low-carbon economic development, and climate change mitigation. More than 27,000 products and services have been carbon labeling worldwide, and more than 12 countries or regions have passed legislation requiring carbon labeling systems [2].

In China, with the introduction of the "double carbon" target and the continuous promotion of carbon labeling in various industries, the discussion on carbon labeling has become more and more enthusiastic, and the application of carbon labeling has become more and more widespread. However, the scope of carbon labeling in China is still relatively small. The government, as a guide, suffers from a deficiency in the formulation and implementation of relevant policies to attract enterprises and consumers to follow, which makes the promotion and application of carbon labeling face great challenges. Therefore, it is crucial to explore how the government can guide enterprises to implement carbon labeling and consumers to purchase carbon label products by formulating effective policies.

The promotion and application of carbon labeling is a complex system involving multiple stakeholders such as government, enterprises, and consumers. At present, this field has been studied by many scholars from the stakeholder perspective. Cohen

C. Liu · X. Wu (✉)
Xiamen University of Technology, Xiamen, China

Z. Peng (ed.), *Trends and Technological Challenges in Green Energy*, Green Energy and Technology, https://doi.org/10.1007/978-3-031-55579-4_11

et al. found that government subsidies can stimulate enterprises to adopt new technologies based on the Stackelberg game [3]. The study by Du et al. using Stackelberg's game shows that consumers' low-carbon preferences can increase enterprises' carbon emission reductions [4]. Tian et al. modeled a game between enterprises and consumers using carbon footprint as a production signal, which revealed that a stable equilibrium between the two is possible only under government policy intervention [5]. Zhao et al. studied the strategy interaction between enterprises and enterprises in the carbon label product market and showed that both direct subsidies and tax incentives have a positive impact on the implementation of carbon labeling schemes by enterprises and that the combination of the two incentives is more effective than using them alone [6]. The strategy interaction between enterprises and consumers was then studied revealing that the introduction of dynamic subsidies has a positive effect on the stability of the market [7]. After that, Zhao et al. studied the behavioral interaction of consumers purchasing carbon label products, and it was detected that product price and price subsidies affect consumers' purchase intention [8].

However, most of the existing studies focus on the behavioral interactions between the two subjects. In fact, government policy guidance, enterprise production decisions, and consumer purchases of carbon label products are often mutually influential, and the game between the two subjects is difficult to accurately portray the decision-making behaviors and evolutionary laws of multiple subjects involving government, enterprises, and consumers at the same time. Moreover, the existing game models are mostly characterized by objective gains and losses, which do not fully reflect the limited rationality of game subjects. The prospect theory can better explain the value perception preference of decision-makers under uncertainty and portray the deviation of decision-making behavior under the influence of subjective judgment [9]. At present, scholars have analyzed the decision-making behavior of game subjects in the game process on prospect theory [10–12], but there is less literature linking prospect theory with the development of carbon labeling in China, especially lack of studies that combine prospect theory with the tripartite evolutionary game.

Accordingly, this chapter introduces the prospect theory to construct the game model of carbon labeling system development from the perspective of stakeholders and adopts the weight function and value function to measure the psychological perceived value of game subjects. The equilibrium point stability of the "government–enterprise–consumer" game system is explored, and the key parameters are simulated and analyzed using MATLAB software. The results of the study provide a theoretical basis and decision reference for promoting the development of China's carbon labeling system and achieving the goal of "double carbon."

2 Model Construction

2.1 Model Assumption

In the process of carbon labeling implementation, the government, enterprises, and consumers, as key stakeholders, constitute a relatively complete system. The tripartite subjects will play repeatedly with the perceived value of benefits and costs.

Hypothesis 1. The government, enterprises, and consumers are all limited rational decision-makers who make decisions in their maximization of interests, and their strategy choices are influenced by their own information and subjective preferences.

Hypothesis 2. Each subject will make its strategy choice based on its own perceived value, which is consistent with prospect theory. According to prospect theory, when facing uncertain gain or loss, game subjects will generate subjectively perceived value. The subjectively perceived value can be reflected by the prospect value V, which is measured by the value function v(Δx) and the weight function ω(p), and the expression of the prospect value is

$$\begin{cases} V_i = \sum v(\Delta x_i)\omega(p_i) \\ v(\Delta x_i) = \begin{cases} \Delta x_i{}^{\alpha}, \Delta x_i \geq 0 \\ -\lambda(-\Delta x_i)^{\alpha}, \Delta x_i < 0 \end{cases} \end{cases} \quad (1)$$

where V represents the game subject's perceived value of gain, Δx_i is the difference between the actual gain of the game subject's choice of strategy and the reference point after the occurrence of the event i, $\Delta x_i \geq 0$ means that the game subject's psychological perception of the behavior is "gain," $\Delta x_i < 0$ means that the psychological perception of the behavior is "loss"; p_i is the probability of the event i, $\omega(p_i)$ is the decision weight, which is a monotonically increasing function, $\omega(0) = 0$, $\omega(1) = 1$; $\alpha \in [0,1]$ is the risk preference coefficient, representing the sensitivity of the game subject to gains and losses, larger α indicates that the game subject is less sensitive to value, the more inclined to take risks; λ is the loss aversion coefficient ($\lambda > 1$), the larger λ is the greater the degree of aversion [9].

Hypothesis 3. Under incomplete information, the set of strategies of the government is {incentive, no incentive}; the set of strategies of the enterprise is {implement, not implement}; the set of strategies of the consumer is {purchase, not purchase}. x represents the probability that the government chooses the incentive strategy, y represents the probability that enterprises choose to implement carbon labeling, z is the probability that consumers choose to purchase carbon label products; and $x, y, z \in (0,1)$.

Hypothesis 4. The government's basic benefit is R_0, and the increase in credibility after choosing the incentive strategy will give it an additional expected benefit of R_1, as well as pay for the related policy promotion cost C_1; for the environmental pollution caused by enterprises not implementing carbon labeling, the government will bear the environmental treatment cost C_2; the government's subsidy coefficient for enterprises implementing carbon labeling is m, and the subsidy coefficient for consumers purchasing carbon label products is n; the carbon tax rate for enterprises implementing carbon labeling is t_0, and the carbon tax rate for enterprises not implementing carbon labeling is t_1.

Hypothesis 5. The product benefit that enterprises can obtain without implementing carbon labeling is R_3, the cost to be paid is C_4, and the carbon emissions generated is e_0; the expected product benefit that enterprises can obtain when they choose to implement carbon labeling is R_2, the cost to be paid is C_3, and the carbon emissions generated is e_1.

Hypothesis 6. When enterprises choose to implement carbon labeling to produce carbon label products and consumers choose to purchase carbon label products, the expected benefit consumers can get is R_4, and the cost they need to pay is R_2; when enterprises do not implement carbon labeling and consumers choose to purchase carbon label products, consumers can only purchase ordinary products, and the cost they need to pay is R_3; the government chooses the incentive strategy to bring environmental benefits is P, and the consumer's low-carbon awareness is μ.

According to the above assumptions, the perceived benefit matrix of strategy choice among the tripartite subjects of the "government–enterprise–consumer" game was constructed, as shown in Table 1.

2.2 Replication Dynamic Equation

By the above model assumptions and variables, let the perceived benefits of the government's "incentive" strategy be U_x, the perceived benefits of the "no incentive" strategy be U_{1-x}, and the average perceived benefits be $\bar{U}_g$, Among them:

$$
\begin{aligned}
U_x = {} & yz[R_0 + v(R_1) - C_1 + v(-mC_3) + v(-nR_2) + v(t_0e_1)] \\
& + y(1-z)[R_0 + v(R_1) - C_1 + v(-mC_3) + v(t_0e_1)] \\
& + (1-y)z[R_0 + v(R_1) - C_1 - C_2 + v(t_1e_0)] \\
& + (1-y)(1-z)[R_0 + v(R_1) - C_1 - C_2 + v(t_1e_0)]
\end{aligned}
$$

$$
U_{1-x} = yzR_0 + y(1-z)R_0 + (1-y)z(R_0 - C_2) + (1-y)(1-z)(R_0 - C_2)
$$

The result of the replication dynamic equation for the government's strategy choice is

Table 1 "Government–enterprise–consumer tripartite" perceived benefits matrix

	Enterprise implement		Enterprise not implement	
	Consumer purchase	Consumers not purchase	Consumer purchase	Consumers not purchase
Government incentive	$R_0 + v(R_1) - C_1 + v(-mC_3)+$ $v(-nR_2) + v(t_0e_1)$	$R_0 + v(R_1) - C_1+$ $v(-mC_3) + v(t_0e_1)$	$R_0 + v(R_1) - C_1 -$ $C_2 + v(t_1e_0)$	$R_0 + v(R_1) - C_1 -$ $C_2 + v(t_1e_0)$
	$v(R_2) - C_3 + v(mC_3)+$ $v(-t_0e_1)$	$-C_3 + v(mC_3)+$ $v(-t_0e_1)$	$R_3 - C_4 + v(-t_1e_0)$	$-C_4 + v(-t_1e_0)$
	$P - R_2 + v(R_4)+$ $v(nR_2) + \mu E$	P	$P - R_3 - S$	$P - S$
Governmentno incentive	R_0	R_0	$R_0 - C_2$	$R_0 - C_2$
	$v(R_2) - C_3$	$-C_3$	$R_3 - C_4$	$-C_4$
	$-R_2 + v(R_4) + \mu E$	0	$-R_3 - S$	$-S$

$$F(x) = \frac{dx}{dt} = x\left(U_x - \overline{U}_g\right) = x(1-x)$$
$$\times \{yzv(-nR_2) + y[v(-mC_3) + v(t_0e_1) - v(t_1e_0)] + v(R_1) + v(t_1e_0) - C_1\} \quad (2)$$

Similarly, the replication dynamic equation for enterprise strategy choice can be as

$$F(y) = \frac{dy}{dt} = y\left(U_y - \overline{U}_e\right) = y(1-y)$$
$$\times \{x[v(mC_3) + v(-t_0e_1) - v(-t_1e_0)] + z[v(R_2) - R_3] + C_4 - C_3\} \quad (3)$$

The replication dynamic equation for consumer strategy choice is

$$F(z) = \frac{dz}{dt} = z\left(U_z - \overline{U}_c\right) = z(1-z)$$
$$\times \{xyv(nR_2) + y[v(R_4) - R_2 + R_3 + \mu E] - R_3\} \quad (4)$$

3 Stability Analysis

Let $F(x) = 0$, $F(y) = 0$, $F(z) = 0$, and find the equilibrium point of the system. Since the asymmetric game requires that the evolutionary stable strategy (ESS) must be a strict Nash equilibrium; this chapter only needs to discuss $E_1(0,0,0)$, $E_2(1,0,0)$, $E_3(0,1,0)$, $E_4(0,0,1)$, $E_5(1,1,0)$, $E_6(1,0,1)$, $E_7(0,1,1)$, $E_8(1,1,1)$ the stability of eight pure strategy equilibrium points.

It is known that the ESS of the system can be obtained by the local stability analysis of the Jacobian matrix [13]. The Jacobian matrix of the system is obtained by calculating the partial derivatives of $F(x)$, $F(y)$, and $F(z)$ with respect to x, y, and z:

$$A = \begin{bmatrix} (1-2x)\left\{\begin{array}{l} yzv(-nR_2) + y\begin{bmatrix} v(-mC_3) + v(t_0e_1) - \\ v(t_1e_0) \end{bmatrix} + \\ v(R_1) + v(t_1e_0) - C_1 \end{array}\right\} & x(1-x)\begin{bmatrix} zv(-nR_2) + v(-mC_3) + \\ v(t_0e_1) - v(t_1e_0) \end{bmatrix} & x(1-x)[yv(-nR_2)] \\ y(1-y)[v(mC_3) + v(-t_0e_1) - v(-t_1e_0)] & (1-2y)\left\{\begin{array}{l} x[v(mC_3) + v(-t_0e_1) - v(-t_1e_0)] + \\ z[v(R_2) - R_3] + C_4 - C_3 \end{array}\right\} & y(1-y)[v(R_2) - R_3] \\ z(1-z)yv(nR_2) & z(1-z)[xv(nR_2) + v(R_4) - R_2 + R_3 + \mu E] & (1-2z)\left\{xyv(nR_2) + y\begin{bmatrix} v(R_4) - R_2 + \\ R_3 + \mu E \end{bmatrix} - R_3\right\} \end{bmatrix}$$

Combined with the Lyapunov indirect method to judge the stability state of the system by the sign of the eigenvalues, if all the eigenvalues of the Jacobian matrix corresponding to the equilibrium point are negative, the point is ESS, otherwise it is a saddle point. The judgment results of the asymptotic stability of the eight pure strategy equilibrium points of the above replication dynamic system are shown in Table 2.

In practice, the initial parameters should meet $C_3 - C_4 > 0$ and $R_3 > 0$. The reasons are as follows: more costly input is required to upgrade technology and equipment as a result of carbon labeling than in the traditional model, and the revenue gained by the enterprise in the traditional model is greater than 0. In this context, $E_3(0,1,0)$, $E_4(0,0,1)$, and $E_6(1,0,1)$ are saddle points, while $E_1(0,0,0)$, $E_2(1,0,0)$, $E_5(1,1,0)$, $E_7(0,1,1)$, and $E_8(1,1,1)$ are possible to establish in each equilibrium case under certain conditions, but this chapter aims to propose the ideal model that promotes the development of carbon labeling, an active implementation by enterprises, and gradual acceptance by consumers, that is, the evolutionary game model eventually evolves to the state of $E_8(1,1,1)$, so only the most ideal stable equilibrium state of $E_8(1,1,1)$ is analyzed. As can be seen from Table 2, in order to reach the ideal state E_8, it is necessary to satisfy $C_1 - v(-nR_2) - v(-mC_3) < v(R_1) + v(t_0e_1)$, $R_3 + v(-t_1e_0) - C_4 < v(R_2) + v(mC_3) + v(-t_0e_1) - C_3$, and $R_2 < v(R_4) + v(nR_2) + \mu E$, that is, the perceived benefits gained from the government's incentive strategy are greater than its costs; moreover, measures such as direct subsidies and collecting carbon taxes can effectively increase the perceived benefits of enterprise implementation and consumer purchases.

4 System Simulation

In this chapter, further numerical simulation analysis is done for the most ideal stable equilibrium state $E_8(1,1,1)$ with the help of MATLAB software to verify the reliability of the model. In order to make the strategy combination evolve to the $E_8(1,1,1)$ ideal state, the parameter values are set to satisfy $C_1 + \lambda(nR_2)^\alpha + \lambda(mC_3)^\alpha < R_1^\alpha + (t_0e_1)^\alpha$, $R_3 - \lambda(t_1e_0)^\alpha - C_4 < R_2^\alpha + (mC_3)^\alpha - (t_0e_1)^\alpha - C_3$, $R_2 < R_4^\alpha + (nR_2)^\alpha + \mu E$. Referring to the assignment methods of scholars Xu [14] and Zhao [15], combined with the realistic situation to assign the parameters. The specific parameters are set as follows: $R_1 = 15$, $C_1 = 2$, $R_2 = 8$, $R_3 = 6$, $C_3 = 5$, $C_4 = 4$, $R_4 = 30$, $u = 1$, $E = 5$, $m = 0.04$, $n = 0.03$, $e_0 = 1.5$, $e_1 = 0.75$, $t_0 = 1$, $t_1 = 1.5$, $q = 2.25$, $b = 0.88$.

Table 2 Eigenvalues and stability conditions of different equilibrium points

Equilibrium points	λ_1	λ_2	λ_3	Equilibrium conditions
$E_1(0,0,0)$	$v(R_1) + v(t_1 e_0) - C_1$	$C_4 - C_3$	$-R_3$	$\lambda_1 < 0, \lambda_2 < 0, \lambda_3 < 0$
$E_2(1,0,0)$	$C_1 - v(R_1) - v(t_1 e_0)$	$v(mC_3) + v(-t_0 e_1) - v(-t_1 e_0) + C_4 - C_3$	$-R_3$	$\lambda_1 < 0, \lambda_2 < 0, \lambda_3 < 0$
$E_3(0,1,0)$	$v(-mC_3) + v(t_0 e_1) + v(R_1) - C_1$	$C_3 - C_4$	$v(R_4) - R_2 + \mu E$	Saddle point
$E_4(0,0,1)$	$v(R_1) + v(t_1 e_0) - C_1$	$v(R_2) - R_3 + C_4 - C_3$	R_3	Saddle point
$E_5(1,1,0)$	$C_1 - v(-mC_3) - v(t_0 e_1) - v(R_1)$	$C_3 - C_4 + v(-t_1 e_0) - v(-t_0 e_1) - v(mC_3)$	$v(nR_2) + v(R_4) - R_2 + \mu E$	$\lambda_1 < 0, \lambda_2 < 0, \lambda_3 < 0$
$E_6(1,0,1)$	$C_1 - v(R_1) - v(t_1 e_0)$	$v(mC_3) + v(-t_0 e_1) - v(-t_1 e_0)$ $+v(R_2) - R_3 + C_4 - C_3$	R_3	Saddle point
$E_7(0,1,1)$	$v(-nR_2) + v(-mC_3)+$ $v(t_0 e_1) + v(R_1) - C_1$	$R_3 - v(R_2) + C_3 - C_4$	$R_2 - v(R_4) - \mu E$	$\lambda_1 < 0, \lambda_2 < 0, \lambda_3 < 0$
$E_8(1,1,1)$	$C_1 - v(-nR_2) - v(-mC_3)$ $- v(t_0 e_1) - v(R_1)$	$v(-t_1 e_0) - v(-t_0 e_1) - v(mC_3)$ $+R_3 - v(R_2) + C_3 - C_4$	$R_2 - v(nR_2) - v(R_4) - \mu E$	$\lambda_1 < 0, \lambda_2 < 0, \lambda_3 < 0$

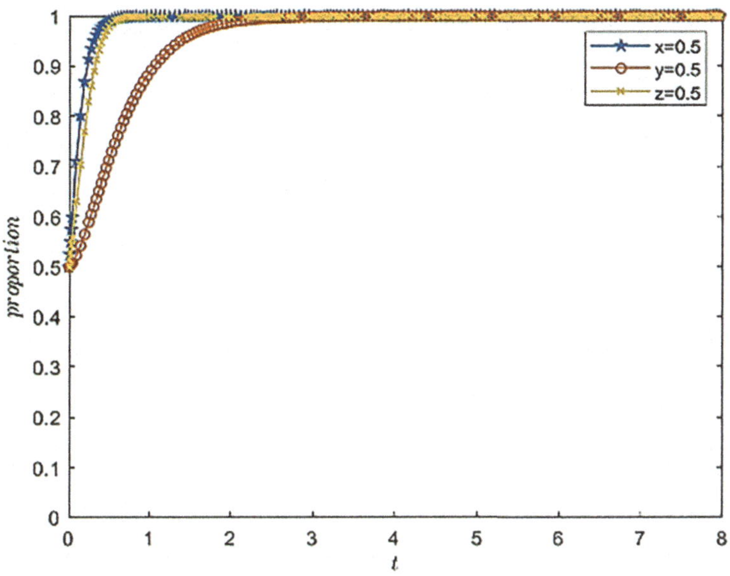

Fig. 1 Tripartite evolutionary stabilization strategy

4.1 Analysis of Stabilization Strategies for Tripartite Evolutionary Game

Through numerical simulation of the above parameters, it can be verified that the government, enterprises, and consumers can evolve to a stable strategy with t under the satisfaction of the set constraints. As shown in Fig. 1, the initial values are $x = 0.5$, $y = 0.5$, and $z = 0.5$. When the above conditions are satisfied, the strategy combination of the tripartite subjects eventually evolves to the $E_8(1,1,1)$ ideal state, when the strategies of the government, enterprise, and the consumer are incentive, implement, and purchase, respectively.

4.2 The Impact of Key Parameters on the Evolution of Behavioral Strategies of Subjects

Impact of Carbon Tax Rate on Government and Enterprise Strategies The impact of the carbon tax rate on government and enterprise strategies is analyzed by changing the value of the carbon tax rate t_1 levied by the government on enterprises that do not implement carbon labeling to 1.2, 1.5, 2.5, 3.5, and 4.5, respectively, while keeping other parameters constant. The simulation results of the evolution of the government strategy are shown in Fig. 2. Regardless of the value of t_1, the

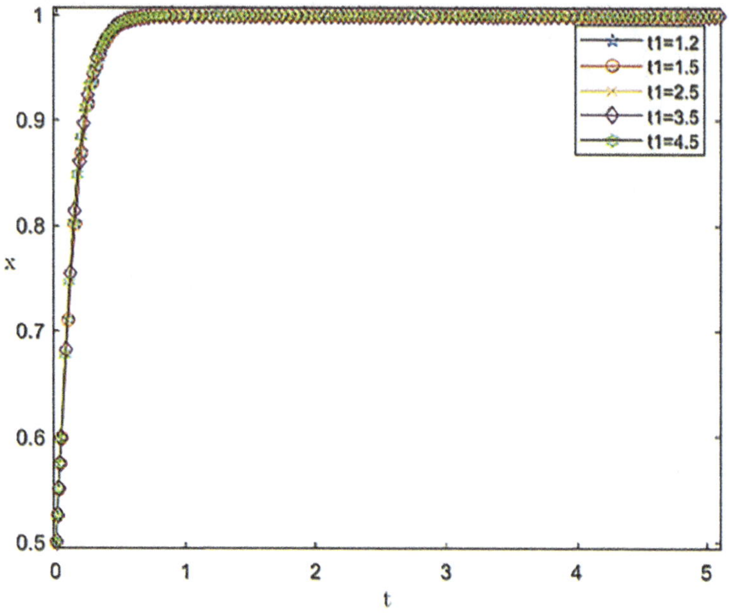

Fig. 2 t_1 on government strategy

simulation curves corresponding to x completely coincide and finally converge at $x = 1$. This indicates that in this system, the government's strategy choice is not affected by t_1, and the government will choose to incentivize regardless of whether the carbon tax rate is large or small. The simulation results of the evolution of the enterprise strategy are shown in Fig. 3. Under the combined factors, although changing only the value of t_1 does not affect the enterprise's evolutionary stabilization strategy, its convergence to implementation will be accelerated with the increase of t_1. Thus, an appropriate increase in t_1 is important to promote the implementation of the enterprise.

Impact of Subsidy Coefficients on Enterprise and Consumer Strategies. Under the premise of keeping other parameters unchanged, the government subsidy coefficient m for the enterprise implementation is assigned and simulated, and the evolutionary path of the enterprise strategy is obtained as shown in Fig. 4. As can be seen from Fig. 4, when m takes the values of 0.01 and 0.04, the corresponding simulation curves converge at $y = 0$, while the simulation curves with m taking values greater than 0.04, the larger the value, the faster it approaches to 1. This demonstrates that in this system m plays a facilitating role in enterprise implementation. The increasing subsidy coefficient causes the evolution of the enterprise's strategy to accelerate. However, too low subsidy coefficients do not promote but rather lead the evolutionary direction toward not implementing.

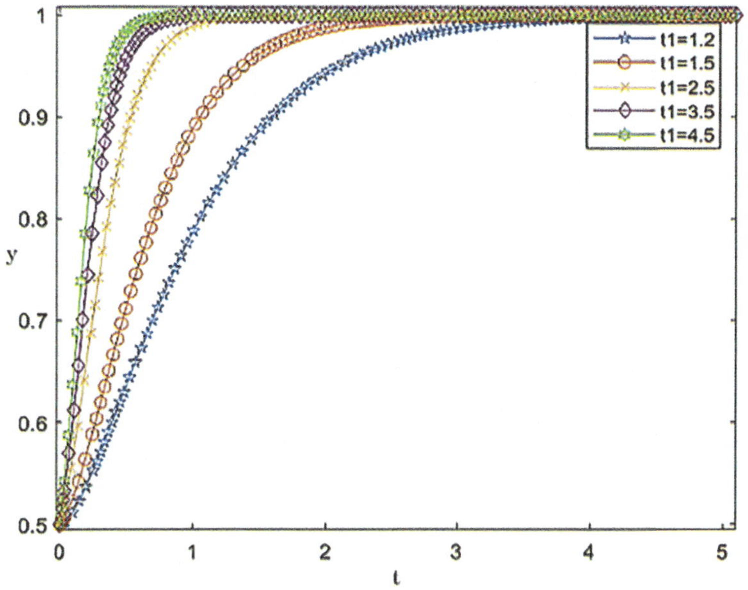

Fig. 3 t_1 on enterprise strategy

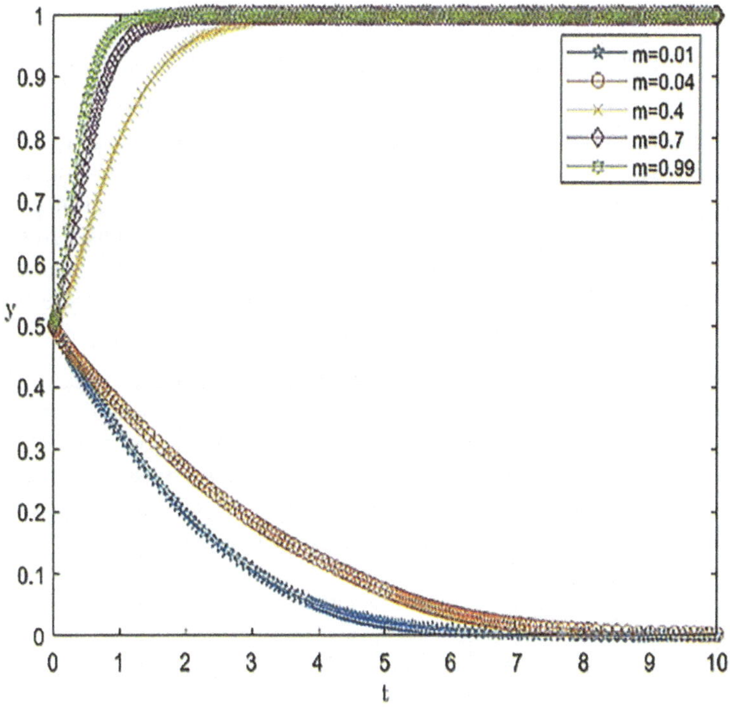

Fig. 4 *m* on enterprise strategy

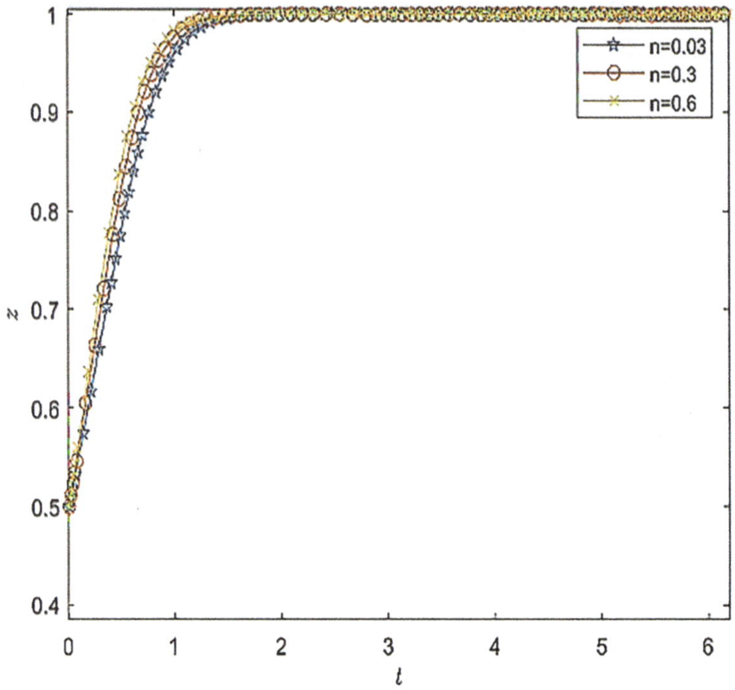

Fig. 5 *n* on consumer strategy

Similarly, the evolutionary path of consumer strategy choice can be obtained in Fig. 5. As the value of n increases, the speed of convergence of the simulation curve to $z = 1$ is accelerated, indicating that an increase in the coefficient of government subsidy to consumers enhances consumers' willingness to purchase. But it can be found that the increasing trend is not obvious, which suggests that consumers are not sensitive to the government subsidy, so if government subsidies are applied to consumer purchases, then consumers will be inclined to purchase.

Impact of Low-Carbon Awareness on Consumer Strategies The evolutionary path of consumer strategy is shown in Fig. 6 when the low-carbon awareness of consumers μ is assigned and simulated. From Fig. 6, it can be found that when μ takes the values of 0.2 and 0.4, the corresponding simulation curve tends to be close to $z = 0$ at the beginning, while z converges to 1 faster and faster as the value of μ increases. It shows that consumers with lower low-carbon awareness will have a tendency not to purchase, as consumers with low-carbon awareness are usually less likely to consider the environmental impact of a product when purchasing it. As consumers' low-carbon awareness increases, their probability of choosing a purchase strategy increases at the same time before convergence.

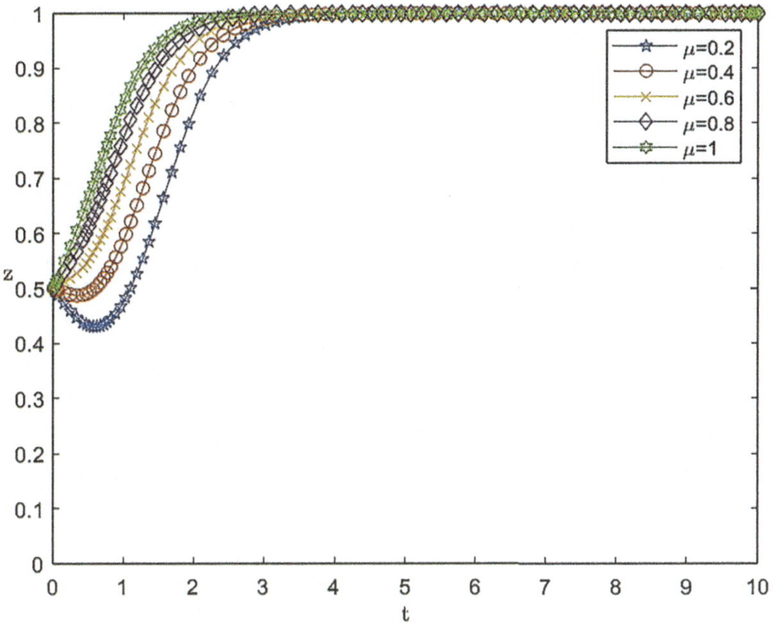

Fig. 6 μ on consumer strategies

5 Conclusions and Suggestions

5.1 Conclusions

Under the background of the carbon labeling system, this chapter combines the prospect theory to establish a benefit perception matrix that is different from the traditional benefit matrix, constructs the "government–enterprise–consumer" evolutionary game model, and analyzes the equilibrium point and stability of the model through the replication dynamic equations. Numerical simulation experiments are carried out with MATLAB software to verify the reliability of the model and analyze the influence of key parameters on the evolutionary trajectory of the tripartite subjects, and the following conclusions are drawn:

1. When the conditions of $C_1 - v(-nR_2) - v(-mC_3) < v(R_1) + v(t_0e_1)$, $R_3 + v(-t_1e_0) - C_4 < v(R_2) + v(mC_3) + v(-t_0e_1) - C_3$ and $R_2 < v(R_4) + v(nR_2) + \mu E$ are satisfied, the strategy combination of the game system will eventually evolve to the $E_8(1,1,1)$ ideal state. At this point, the perceived additional expected benefits of government incentives with carbon tax revenues are greater than the sum of the subsidies and associated promotional costs, the enterprise's profit from implementation is greater than the profit from nonimplementation, the perceived benefits and subsidies from consumer purchases together with the total environmental and ecological value of the purchase are greater than the purchase costs.

2. Government sensitivity to carbon tax rates is low, with a preference for "incentive" regardless of whether the carbon tax rate is high or low.
3. For enterprises, both government subsidies and carbon tax rates can effectively incentivize implementation. Among them, government subsidies have a threshold effect, and an increase in the subsidy coefficient within a certain range can increase the enthusiasm of enterprises to implement, but too low a subsidy coefficient is difficult to motivate enterprises to implement. In addition, enterprises are more sensitive to government subsidies than consumers.
4. For consumers, government subsidies and low-carbon awareness can prompt the purchase of carbon label products.

5.2 *Suggestions*

In summary, to promote the development of carbon labeling, this chapter proposes the following recommendations:

1. The government, as the leader in promoting the development of carbon labeling, should fully play its role. While formulating incentive policies, it should take current emerging technologies such as the Internet of Things and artificial intelligence into account, to reduce the publicity cost of relevant policies, improve the return rate, and fully utilize the incentive role of the government.
2. Government subsidies are an effective measure to motivate enterprises and consumers to choose a carbon labeling strategy. Therefore, at the early stage of carbon labeling development, the government should design scientific and reasonable subsidies to support enterprises and consumers, trying to maximize the incentive effect with minimum input. In addition, setting different carbon tax rates for enterprises with or without carbon labeling can also motivate them to implement carbon labeling.
3. The government can continue to expand low-carbon consumption scenarios, covering not only superstores, tourist attractions, restaurants, and hotels, but also transportation hubs such as airports and train stations.
4. Enterprises can improve the package design of carbon label products to show the green product information concisely and effectively so that consumers can more intuitively understand the social and environmental benefits that carbon label products can bring and increase the perceived benefits of purchasing.
5. Consumers should enhance their low-carbon awareness, actively respond to the government's appeal for low-carbon consumption, and give more consideration to the greenness of products.

Funding The Social Science Foundation of Fujian Province (FJ2023B061), Fujian Province Innovation Strategy (Soft Science) Research Project (2023R0108), Xiamen Social Science Research Project (Key Project) are funded by the Xiamen Federation of Social Sciences and Xiamen Academy of Social Sciences (Xiamen Social Science Research [2023] B12).

References

1. Anton B, Cambray A, Dupar M, Mairi Dupar, W Astrid, Gogoiet E. Close to home: subnational strategies for climate compatible development[J]. The Climate and Development Knowledge Network (CDKN), 2014:1–21
2. CARBON TRUST. Footprint label-certified organisations[EB/OL]. (2022-08-31) [2022-10-06].https://www.carbontrust.com/whatwe-do/assurance-and-certification/productcarbon-footprint-label/footprint-label-certified.
3. Cohen M C, Lobel R, Perakis G. The impact of demand uncertainty on consumer subsidies for green technology adoption[J]. Management Science, 2016, 62(4):1235–1258.
4. Du S, Zhu J, Jiao H. Game-theoretical analysis for supply chain with consumer preference to low-carbon[J]. International Journal of Production Research, 2015, 53:3753–3768.
5. Tian Z Y, Yuan R P, Huo L Y. Pricing, carbon footprint and signaling game[C]. 2017 2nd International Conference on Civil, Transportation and Environmental Engineering (ICCTE 2017). 2017, 505–512.
6. Zhao R, Zhou X, Han J, Liu C. For the sustainable performance of the carbon reduction labeling policies under an evolutionary game simulation [J]. Technological Forecasting and Social Change, 2016, 112: 262–274.
7. Zhao R, Han J, Zhong S, Huang Y. Interaction between enterprises and consumers in a market of carbon label products: a game theoretical analysis[J]. Environmental Science and Pollution Research, 2018, 25(2):1394–1404.
8. Zhao R, Zhou X, Min XEF. Game analysis of carbon label product purchase behavior based on system dynamics [J]. Journal of Systems Science, 2021, 29(02):111–115.
9. Kahneman D, Tversky A. Prospect theory: ananalysis of decision under risk[J]. Econometrica, 1979, 47(2):263–291.
10. HE X Y. Research on the evolution mechanism of the electric vehicle market driven by big data [J]. Concurrency and Computation: Practice and Experience, 2020, 32(6):1–12.
11. Zhao Z, Man Q. Game analysis on the evolution of risk management behavior of major infrastructure projects based on prospect theory [J]. Journal of Systems Management, 2018, 27(01):109–117.
12. Wang X, Zhang L, Zhang N. A three-way game analysis of health information quality of WeChat based on prospect theory [J]. Computer Science, 2022, 49(S1):694–704.
13. Friedman D. On economic applications of evolutionary game theory[J]. Journal of Evolutionary Economics, 1998, 8(1):15–43.
14. Xu J, Xu Y. Diffusion mechanism of low-carbon technology innovation under government environmental regulation-an evolutionary game analysis based on prospect theory [J]. Systems Engineering, 2015, 33(2):118–125.
15. Zhao Z, Liu Y, Liu L. Evolutionary game analysis of drug quality and safety regulation based on prospect theory [J]. Operations Research and Management Science, 2022, 31(10):75–81.

References

Index

B
Buck–boost bidirectional power converter, 4
Building integrated photovoltaics (BIPV), 13, 16, 17, 20–22, 102

C
Carbon labeling, 125–128, 131, 133, 137, 138
CFD, 114, 120
Climate change, 12–15, 19, 37, 38, 59, 125
Combustion, 13, 14, 16, 87, 113–122
Conventional proportional integral, 40

D
DC/DC power converters, 3–9
Dynamic monitoring, 53

E
Evolutionary game, 125–138
Extreme wind conditions, 73–80

F
Finite element analysis, 75, 77

G
Generative deep learning, 60, 61, 68

H
Hybrid renewable energy system (HRES), 99–110
Hydrogen storage, 100, 102, 103, 105, 110

I
Interferometric radar, 49–56

L
Large span valve hall, 73–80
Light-emitting diodes (LEDs), 37–44
Luo based bidirectional power converter, 9
Luo converters, 4–9

M
Machine learning, 59–61
Multi-objective optimization, 106

N
Nature-based solutions (NBS), 13, 16, 18, 19, 21, 22
Nondestructive testing (NDT), 49, 50

O
Ocean power generators, 60, 61, 68
Operational neural networks (ONNs), 50

Z. Peng (ed.), *Trends and Technological Challenges in Green Energy*, Green Energy and Technology, https://doi.org/10.1007/978-3-031-55579-4

P
Power factor correction (PFC), 38–44
Prospect theory, 125–138

R
Regenerative cooling, 113–122
Renewable energy, 4, 14, 15, 21, 27–29, 34, 59, 69, 85–96, 99, 103, 108, 110

S
Sheppard-Taylor converter (STC), 40, 43
Simulation analysis, 131
Simulations, 3–9, 21, 80, 86, 90–96, 107, 113–122, 133, 134, 136, 137
Solar incentives, 34
Solar subsidy, 31, 33
System dynamics, 86, 88–90, 96

T
Tax credits, 27, 29
Tax rebate, 28
Tidal power generator, 59–69
Total harmonic distortion (THD), 39, 40, 43, 44

V
Variable auto encoder (VAE), 59–69

W
Wind turbine blades, 49–56

SPRINGER NATURE

GPSR Compliance

The European Union's (EU) General Product Safety Regulation (GPSR) is a set of rules that requires consumer products to be safe and our obligations to ensure this.

If you have any concerns about our products, you can contact us on ProductSafety@springernature.com

In case Publisher is established outside the EU, the EU authorized representative is:

Springer Nature Customer Service Center GmbH
Europaplatz 3
69115 Heidelberg, Germany

The manufacturer's authorised representative in the EU is Springer Nature Customer Service Centre GmbH, Europaplatz 3, 69115 Heidelberg, Germany. If you have any concerns regarding our products, please contact ProductSafety@springernature.com

Printed and bound by CPI Group (UK) Ltd, Croydon, CR0 4YY
01/10/2026
02225916-0009